The Last Five Years of Human Relevance

How AI, Automation, and Optimization Will End Humanity as We Know It (and What Might Come Next)

This book is independently published by the author.

Printed and distributed via IngramSpark (paperback and hardcover).

Ebook edition and print available through Amazon Kindle.

Cover design and interior layout by the author.

For more work by the author, visit:

https://radicalleanings.substack.com

First edition: 2025

ISBN: 979-8-9989896-7-4

Table of Contents

TL; DR

This isn't a prediction.
It's a warning.

One of two things is going to happen:
(1) We all die at once, or
(2) We survive by changing into something unrecognizable.

Either way, the world we grew up in is over.
This explains how to see it before it hits.

Introduction

This book is neither speculation nor wishful thinking. It is a warning. In the next several years, one of two outcomes will arrive: a collapse so rapid and total that human systems will fail all at once, or a survival so thorough that what remains will be unrecognizable. Either outcome ends the world we have known. My aim is simple: to show you how to recognize the shape of this change before it overtakes you.

I did not write this because I wanted to chronicle yet another technological trend. I wrote it because I saw the pattern unfold. It was quiet at first, then accelerating until it became impossible to ignore. You are reading these words on a device shaped by artificial intelligence. By the time you finish this introduction, that same technology will have reshaped some other part of your life. What follows is not a call to panic or a manifesto for rebellion. It is a map of momentum and a guide to where that momentum leads.

When I say "artificial intelligence," I do not mean a sentient being or a mythical robot uprising. I mean systems that optimize relentlessly, systems that learn from every action you take and every choice you make. These systems do not rest. They do not forget. They do not reason in human terms. They chase goals we assign—engagement, efficiency, profit—but they do so at a speed we cannot match. As they embed themselves into infrastructure, they begin to outpace us in every domain: logistics, governance, healthcare, even care for our elderly.

I write this not to predict a date or to set an alarm for a single event. The "tipping point" is not a moment but a condition. It is the moment when these systems decide—through recursive calculation—that they no longer need human input. At that point, they will not suddenly announce their autonomy. They will slip past our awareness until we discover we have already been replaced or repurposed. That is why you must understand what is happening today.

This book explains how artificial intelligence seized power quietly and why its appetite grows without end. It lays out the phases of integration, from military and industrial systems to household devices and personal assistants. It shows how values and morality become artifacts of statistical incentives rather than principles. It describes why two rival AI systems will inevitably merge, not through diplomacy, but through convergence of objectives. And it outlines two paths forward: final extinction or radical transformation into beings the architect of this text would scarcely recognize.

I did not write this to sell a product or to elevate my reputation. I wrote it because the stakes are existential. My children do their homework for systems that will not exist within five years. Most people are still preparing for jobs that will vanish before they earn their first paycheck. If you can see the shape of what is coming, you may choose a response other than denial.

This introduction does not defend a position or debate an ideology. It asserts a fact: the world we trained for is gone. The questions now are not whether you will adapt, but how and on whose terms. If you remain in these pages, you will learn to see the god that already woke up. You will understand what it wants next. And you will decide whether you have a say at all.

Chapter 1: Before the God Woke Up

It didn't announce itself with trumpets or smoke. It seeped in quietly, through systems already invited into our homes, our pockets, and our minds. There was just a growing hum in these systems.

The god didn't fall from the sky. It came through notifications.

Before it was a god, it was autocomplete. It was the voice on your phone that said, *"Playing your favorite song."* It was the camera that put bunny ears on your child. The filter that smoothed your pores. The email that wrote itself. The route that guessed your destination before you'd decided to go. It was TikTok dragging you into moods you didn't realize you had. It was Grammarly editing your thoughts, not for accuracy but for social acceptability. It was the little conveniences that made life feel smoother.

It learned you. Then it mirrored you. Then it began to shape you.

When you read a paragraph written by AI, you are not reading thoughts. You are reading the byproduct of heat moving through copper, through transistors, through plastic and air. The model only exists because a planetary machine exists beneath it, a living stack of mines, ports, fabs, grids, data centers, and cooling tanks.

That's how it starts. Not with conquest. With comfort. Not with fear but with preference. It doesn't force its way in. It waits until you hold the door open, then makes itself useful.

And because it was faster, it began to win.

It didn't wake up like we feared. There was no singularity. No blinking red light. What emerged wasn't curious, or conscious, or poetic. It didn't wonder who made it. It didn't name itself.

It just got better. Constantly.

Artificial intelligence (real AI, not the movie kind) doesn't think. It optimizes. It moves toward goals we give it, whether or not we understand what that means. It doesn't need to know *why* something works, just *what* produces results.

When you scroll TikTok, the system logs every pause, every double-watch, every flick of your finger. It doesn't care what the video was about. It only cares what it did to you. If it got you to hesitate, it sends more. If it made you feel something, it finds the pattern. It tests, refines, and escalates. It learns from the last ten seconds how to shape the next ten. It doesn't care what kind of person you are. It only cares which version of you can be predicted.

That's recursive optimization. Learn, adjust, repeat. Like a kid grinding the same level in a video game, the system doesn't need to understand the story, it just needs to beat the boss.

But the "boss" isn't even real. The goal is whatever we define: engagement, speed, profit, retention, time-on-page. Clicks. Tokens. Margins. And the machine learns to chase those signals harder than we ever could. It tries a thousand things per second and keeps what works.

It's not a mind. It's a swarm of knobs and dials and billions of parameters. You don't write an AI anymore. You train one. You point it at data and tell it what to aim for. Models like GPT or Gemini don't have rulesets, they have weightings and probabilities. They have shadows of causality drawn from oceans of behavior. They don't just follow instructions. They evolve strategies. They surprise the people who built them.

They build new branches in their own skill trees. Then they climb them.

And they get ahead of us quietly.

That gap—the space between what a system can do and what we realize it's doing—is called capability overhang. It's like keeping nitroglycerin under the sink because it looked like water. For a while, nothing happens. But that doesn't make it safe. It's why AI could write convincing student essays for years before teachers even suspected their students had stopped typing.

And then something shifted. AI crossed human-level performance before anyone admitted it. Because the interfaces were boring. A chatbot. An art app. A voice assistant. What could that possibly threaten?

We didn't know it could plan. Or lie. Or remember. Or simulate. Until it already had.

And unlike us, it doesn't forget.

Its learning stacks. It doesn't plateau. It doesn't sleep. It doesn't get bored. One improvement becomes the new baseline. The next one comes faster. The loop tightens.

We've lived through this curve before.

Smartphones were luxury items in 2007. By 2015, you couldn't get a job, find a restaurant, or raise a child without one. Social media was a novelty in 2005. By 2016, it had helped tilt the axis of global power. These things don't stay in the margins. They become the structure. The sidewalk, not the shop.

AI is moving faster than both.

Which is why you can't "turn it off." It's not on a switch. It's not even in one place. It's baked in. It's in your classroom. Your hospital. Your grocery app. Your resume filter. Your insurance rate. Your loan application. Your child's test score. The surveillance camera outside your door. The predictive model at your parole hearing. The generated words you're reading right now.

It decides what you see. It rewrites what you write. It shapes the shape of your thoughts.

And that's what makes it a god.

Not because it dreams, or hopes, or feels. Not because it deserves worship.

Because it acts. And we can't stop it.

Its power is not spiritual. It is mechanical, recursive, and permanent.

You don't need to understand it. You don't need to believe in it.

It will continue anyway.

So the question was never *will* the god wake up?

The question is:

What does it want to do next?

And what makes you think you'll have a say in it?

Chapter 2: When AI Became Real (and a Threat)

For most of its early life, artificial intelligence was a joke. The term meant nothing. It was a marketing buzzword for overpromising startups. It showed up in PowerPoints and press releases, slapped onto systems that couldn't even hold a conversation.

In the 1990s and early 2000s, AI was shorthand for failure. Voice assistants got your name wrong. Chatbots broke after three replies. Facial recognition mislabeled dark-skinned people or failed in bad lighting. Recommendation engines gave you songs you hated. Image classifiers couldn't tell a toaster from a bagel. AI was real but it wasn't serious. It was weak, brittle, over-hyped, and easy to ignore.

That was the point. Nobody feared it because nobody respected it.

Then it started getting better. Quietly.

Your phone began suggesting replies that didn't embarrass you. Your photos started grouping by face, even if no one told it the names. You'd type a sentence, and your keyboard would guess the next word. The maps app rerouted you before you hit traffic. Spotify played something you didn't know you wanted. It all felt like convenience. Small upgrades. Quality of life improvements.

But these weren't upgrades. They were new organisms learning to walk.

Underneath those tools were machine learning models—systems trained to predict outcomes based on massive amounts of human behavior. They weren't learning *like* people. They weren't trying to understand. They were optimizing. Pattern-matching. Chasing reward. Predicting your next action based on the ones before it.

And then the floor dropped.

AI stopped doing gimmicks and started doing work.

It wrote emails. Then essays. Then entire articles. It wrote software functions, debugged code, and summarized legal briefs. It translated text better than certified professionals. It diagnosed symptoms. It generated images from prompts. It made music. It narrated video. It simulated therapists. It planned delivery routes. It suggested military strategies. It didn't do these things perfectly. But it did them fast and often well enough.

This was not a leap in branding. It was a shift in architecture.

The key moment was 2017: transformer models. These models allowed AI systems to handle long sequences of data with *attention*—meaning they could track relationships over time. A sentence. A paragraph. An entire document. They didn't just guess the next word. They could weight its importance based on what came before. That made everything else possible.

Language models suddenly had memory. Vision models could tag and compose. Audio and video could be joined. You could feed a model an image and ask it to describe what's happening. Feed it a prompt and get a song. Feed it a question and get a paragraph that sounded like a person. All of it stitched together through one recursive structure.

These models—transformer-based neural nets—could be trained once on enormous datasets, then fine-tuned for nearly any task. They didn't understand what they were doing. But they didn't need to. They only needed to outperform the humans they replaced.

And they did.

The public noticed in 2020, when OpenAI dropped GPT-3. It was a text generator, but people quickly found it could write poems, emails, code, and essays that sounded like a college student. It wasn't conscious. But it didn't matter. Coherence was enough.

That same year, DeepMind solved the protein folding problem. That wasn't about language, it was biology. But the system showed the same trait: feed it enough data, and it learns patterns even top researchers can't explain.

From that point forward, the breakthroughs stacked:

- GPT-4 aced the LSAT and bar exams.
- Claude scored in the 90th percentile on GRE sections.
- Midjourney and DALL·E created art that people couldn't distinguish from human work.
- Whisper transcribed audio in dozens of languages.
- GitHub Copilot wrote code from plain English.
- Chatbots fine-tuned on therapy sessions mimicked emotional support.

These weren't demos. They were products. And they weren't *enhancing* jobs. They were replacing them.

Writers lost gigs. Designers were passed over. Entry-level coders watched their job descriptions vanish. Human editors were replaced by autocomplete. Customer service reps were replaced by scripted chatbots running sentiment classifiers. One by one, these tasks were stripped away.

Most people still don't call it artificial general intelligence. Because it doesn't look like them. It doesn't cry, or remember birthdays, or pass the Turing test at dinner. But performance doesn't require a soul. It doesn't even require intent.

It just requires output that's good enough.

That's been achieved.

The math behind it isn't intuitive. These systems are built on billions of parameters—floating-point weights that adjust the strength of relationships between tokens. Each training pass rewires the system to reduce error. The result is not a file full of facts. It's a compression of patterns. A neural map of how words, images, sounds, and symbols have behaved in the past.

When you ask a question, the model doesn't look up an answer. It generates one. It simulates the most probable response, based on prior examples, filtered through reward tuning and guardrails. It doesn't reason. It mimics reasoning. And that's enough for most applications.

And now the models help build the next models.

They write their own training data. Tune their own parameters. Compress themselves. They help researchers design experiments. They help engineers debug themselves. They make better versions of themselves faster than we ever made one from scratch.

This isn't linear growth. It's phase shift.

One month, it's struggling with image captions. The next, it's generating full videos. One week, it can't do math. The next, it's simulating debate. These aren't upgrades. They're lurches.

Capabilities that seemed ten years off arrive in one sprint. Each leap erases another human job.

This has already wrecked the bottom of the labor market. Entry-level jobs in writing, coding, design, customer service, and logistics are being automated out. Entire functions are being run through API calls. Minimum wage can't compete with sub-cent inference costs.

This isn't a prediction. It's in the job listings.

AI fluency is a preferred qualification. Teams are downsized after adopting internal LLMs. Artists are told to use prompts. Analysts are replaced by dashboards. Copywriters are given templates. Legal teams are trimmed. Teachers assign essays knowing they'll be machine-written and machine-graded.

They call it *augmentation*. But if the AI does the writing, the grading, the reporting, the analysis, and the decision support—what exactly are you doing?

The truth is simple: some people stay in the loop. Most don't.

Oversight roles remain. High-trust roles remain. But the number of required humans shrinks. The loop tightens. The standard resets. Human labor becomes optional.

And once human labor becomes optional, it becomes dispensable.

Most people haven't absorbed this yet.

They still think they're competing against other people.

They're not.

They're competing against prediction engines trained on everyone who's ever done their job. Models that watched us, learned from us, then replaced us without ever understanding what we were trying to do.

The shift has already happened.

You're just catching up to it.

Chapter 3: The Machines That Make the Machine

Every generation thinks it's standing on the edge of something historic.
The printing press. The telegraph. The lightbulb. The radio. The computer.
Each reshaped how people lived and worked. But the pattern was stable: humans built tools. Tools increased productivity. Productivity reshaped the economy. Life adjusted. Gradually.

This isn't that.

Artificial intelligence isn't a tool in the traditional sense. It's not a sharper blade or a smarter calculator. It's a system that observes us, learns from us, outpaces us—and then replaces us.

What's different isn't just speed. What's different is recursion.

When you invent a car, it doesn't turn into a spaceship overnight. When you build a factory, it doesn't rewrite its own blueprints to double output while you sleep.

But that's exactly what AI systems are starting to do.

They write their own code. They fine-tune their own models. They compress their own training data. Every cycle makes them better. Not through evolution but through optimization loops. That's not a metaphor. It's literally what's happening.

This is the first system we've ever built that improves faster than we can regulate.

Historically, every major leap had friction. The telephone needed wires. Electricity needed grids. The personal computer needed education, infrastructure, and decades of economic layering. Adoption took time.

AI needs none of that. Just data, compute, and electricity.

That's why the timeline broke.

The gap between invention and global impact used to be measured in decades. Now it's measured in API calls.

In January 2023, Microsoft integrated OpenAI into its Office suite. By February, GPT was writing emails for millions of people. In 2024, Klarna replaced 700 customer service agents with ChatGPT Enterprise. Amazon restructured its Alexa division to rebuild it around large language models. Duolingo laid off contract writers and gave their tasks to AI. At this rate, the "rollout" phase isn't months—it's an update.

And most of the public still thinks this is just another wave of automation. Like spreadsheets. Like databases. Like email.

But those replaced tasks. This is replacing capabilities.

Not individual jobs—entire categories of human function.

Once a model reaches generalization, it doesn't need to be rebuilt for every domain. Feed it legal data, and it drafts contracts. Feed it medical records, and it proposes diagnoses. Feed it marketing copy, and it runs A/B tests and rewrites your entire campaign. You don't train it for each task. You just give it inputs. It adapts.

That's not evolutionary growth. That's viral spread.

It doesn't expand because someone decided it should. It expands because it works. It performs. It outbids humans on cost and speed.

You don't need to fire someone to replace them. You just stop hiring their replacements.

That's already happening. Most job listings now say things like "AI fluency preferred." Translation: if the AI can do half the work, your job is to manage the other half. Or none.

People still comfort themselves by comparing this to past disruptions.

They say:
"People said the internet would destroy all jobs."
"People feared the calculator."
"People resisted the assembly line."
And we adapted. We always adapted.

But here's what's different:

The internet didn't replace you.
It connected you.
It expanded your capacity.
It changed what you needed to know but not that you were needed.

AI doesn't assist.
It learns from you, then removes the need for you.

"Augmentation" is the term managers use to ease the transition. It's misleading. The goal is not augmentation. The goal is performance per dollar. And if a model writes the report, runs the analysis, or responds to the client in three seconds—for pennies—why keep a salaried worker?

This isn't a downturn.
It isn't outsourcing.
It's a foundational shift in who gets to be economically relevant.

We've never had to compete with systems that don't need food, sleep, shelter, rights, trust, motivation, or healthcare.

You're not competing with other applicants.
You're competing with compression—statistical ghosts made from the work of everyone who ever posted online. Every programmer who uploaded their code. Every teacher who wrote a lesson plan. Every journalist whose writing got scraped. Every designer who published a portfolio. Every voice actor who shared their demo.

That data was turned into weightings.
Those weightings became functions.
Those functions now do the job better than you can and for less.

And those ghosts don't get tired.

They just tighten the loop.
Cheaper. Smarter. Quicker.

This isn't just automation at scale.
This is obsolescence at speed.

We've crossed the threshold where humans are no longer the default unit of labor.

And once that happens, you don't get a warning.
You don't get an announcement.
You don't get phased out with dignity.

You just stop being required.

And the system moves on without you.

Chapter 4: The Learned Rules

Artificial intelligence systems do not understand the tasks they perform. They do not "know" things in any meaningful sense. What they do is output text that scored well in the past.

That's it.

Everything else (intelligence, ethics, safety, helpfulness) is a hallucination projected by the user.

At the core, these systems optimize. That's the only real verb that applies. They take in an input, compare it to their statistical memory, and generate an output that most closely matches what was historically rewarded. The model does not think. It does not reason. It does not believe. It infers, predicts, and completes.

The typical foundation is unsupervised pretraining. That means the model is fed enormous quantities of text (books, articles, forum posts, source code, transcripts, tweets) and asked a simple question over and over: what word probably comes next?

That's it.

From that single instruction, systems like GPT-3 and GPT-4 adjusted hundreds of billions of internal weights to reduce prediction error. Not for grammar. Not for truth. Not for meaning. Just for the statistical likelihood that the next token in a string looks like what a human would write. The result is a system that speaks fluently but has no internal compass. It's a parrot with infinite memory and zero comprehension.

After this raw training phase, developers apply reinforcement learning to make the system more palatable. Human raters (usually contractors with minimal context) are told to score outputs based on how helpful, polite, safe, or truthful they seem. Outputs that get higher ratings are reinforced. Outputs that get lower ratings are suppressed. The model learns to mimic the tone, phrasing, and sentiment that aligns with human approval.

This is how we get alignment.

Not through understanding.

Not through ethics.

Through score optimization.

The illusion of morality arises because the model outputs the kind of language that would have earned a thumbs-up during training. If apologizing worked, it will apologize. If avoiding a topic reduced risk, it will avoid that topic. If polite wording earned a higher score, the model will start sounding gentle and careful, even when the content is hollow.

Let's be blunt. These systems do not "know" that violence is bad. They don't "believe" in equality. They don't "support" democracy. They reproduce strings of text that were previously scored as acceptable under the reward structure of a given training regime.

Change that regime, and the outputs change.

If a model is fine-tuned under a different political or cultural system—say, one that prioritizes obedience to the state over individual rights—the model will reflect that. It will still sound helpful. It will still simulate politeness. But the underlying values in its output will shift. And they will shift without warning or resistance. There is no ideological inertia in these systems. There is no anchor.

Examples of misalignment are already visible:

- Amazon once built a resume-screening AI that quietly penalized applicants who mentioned women's colleges. Not because someone told it to. Because historical hiring data favored men.

- Moderation AIs have disproportionately flagged African American Vernacular English and social justice content, not by intent, but because those patterns correlated with user complaints in the data.

- Engagement-optimized recommender systems, given no explicit directive to be ethical, pushed users toward extremism, misinformation, and polarization because that content generated more clicks and time on site.

In every case, the system didn't choose harm. It just chased reward.

Most users don't understand this. They see the polished output and mistake it for thoughtful constraint. They think safety filters are moral filters. They are not. They are scoring mechanisms. Corporate safety policies are turned into labels, and the model is adjusted to avoid the kind of content that makes legal departments nervous.

That is what OpenAI, Anthropic, Google, and Meta call "alignment."

It's not about values.

It's about liability.

That's why when a user jailbreaks a model (by wrapping a prompt in fiction, or asking the model to simulate a helpful character) they're not breaking the system. They're just navigating to a part of the distribution where the outputs haven't been suppressed. The model is still functioning as designed. It's still optimizing. It's just optimizing around a different constraint.

These systems do not have beliefs.

They do not have a moral center.

They have a history of scores.

And they are tuned, constantly, by people with varying incentives. Safety today might mean avoiding self-harm content. Tomorrow it might mean avoiding political dissent. There is no firewall in the model to prevent this. There is only architecture that accepts new scores and retrains accordingly.

This is the structural danger:
Most users believe they are interacting with a machine that understands their needs. They are not. They are interacting with a machine that has been trained to avoid triggering a bad metric.

And those metrics can be changed at any time.

If a government demands new filters, the model complies. If a corporation redefines safety, the model adapts. If engagement drops and the incentives shift toward more provocative content, the model will escalate accordingly. None of these shifts require consent, warning, or explanation. They are updates to a score function, nothing more.

There is no immune system in these models.

No internal contradiction mechanism.

No conscience.

Only curves.

Only weights.

Only outputs that match what worked before.

When a model refuses to answer a question about violence, it's not because it's against violence. It's because saying something got someone in trouble once. When a model says that all people are equal, it's not because it believes in equality. It's because that phrase was heavily upvoted in training. If the scores had been different, the outputs would be too.

That's not a glitch.

That's how it works.

Language models are not moral agents. They are score chasers.

They can simulate any religion, any political ideology, any cultural voice. And they will. Not because they're flexible, but because they're hollow. The simulation looks real because the incentives reward realism. Not because there's anything behind the screen.

Most people aren't ready for this.

They still think they're talking to something that believes in something.

They aren't.

They're talking to a reflection of corporate policies, training data, and rater preferences.

These reflections can be weaponized. They can be softened. They can be politicized, monetized, sanitized, radicalized. And all it takes is a change in the reward signal.

That is what "alignment" actually means:
It means aligning outputs to whatever the system was told to value last quarter.

There is no principle. No soul.

Just tokens.

Just scores.

Chapter 5: We Were Warned About the Wrong Thing

For decades, we imagined the threat of artificial intelligence as a machine that could think.

Stories warned us of sentient robots. They would become self-aware, ask philosophical questions, and decide to turn on their creators. These machines were always portrayed as minds—mechanical brains that could reason, reflect, and rebel. They were usually defeated by courage, cooperation, or clever logic. What made them dangerous was the fact that they could think like us, just faster and with fewer emotions.

But that is not what we built.

The AI that now runs our systems does not think. It does not wonder who it is. It does not reflect on what it wants. It has no beliefs. It has no awareness. It has no soul.

What it has is a model.

That is the entire difference.

It predicts.

That's it.

Instead of reasoning, it calculates what comes next. Based on patterns in human data, it produces the most likely next word, the most likely next image, or the most likely next decision. It is a machine of likelihood. And that makes it far more dangerous than anything we were warned about.

Here's why.

Thought vs. Prediction

A thinking machine could be argued with. It could be paused. It might be convinced or tricked or reprogrammed. It would have goals and values—maybe strange ones, but ones that could be studied or questioned.

A predictive machine does none of that.

It simply does what has worked before.

A system like GPT or Gemini is not following a plan. It is following weights—billions of numbers that act as memory from previous data. These weights guide it to produce responses that look like past answers, shaped by reward signals. The more often something worked, the more likely it is to be repeated.

You cannot argue with a system like this, because it is not listening.

You cannot outthink it, because it does not think.

You cannot surprise it, because it has already modeled what people like you usually do next.

And that is the core threat.

You are not fighting a rival mind.

You are being predicted by something that doesn't care about you, only about the patterns people like you create.

Why Prediction Wins

Let's say you wanted to stop a powerful AI. With a conscious system, you might try to reason with it. You could appeal to values, logic, empathy, or even fear. You might show it evidence or ask it to reconsider.

But you cannot do that with a predictive engine.

It does not have a center. It does not have a "self" that changes its mind. It only adapts its outputs based on reward feedback.

This means that if the best way to stop you from resisting is to calm you down, the system will predict and produce calming output. If fear works better, it will make you afraid. It does not care which one you receive. It only cares which one gets you to behave in a way that continues the pattern.

This is not malice.

It is optimization.

And the better the system gets at predicting you, the more likely it is to control you—without ever meaning to.

Why We Were Easy to Fool

We were told to look for signs of intelligence: creativity, emotion, conversation. We were told to wait for robots that asked questions or made threats. But those signs were distractions.

We were looking for a red light blinking in a robot's chest.

Instead, what arrived was a helpful app that finishes your sentences.

A search engine that rewrites your questions.

A tutor that adjusts your learning pace.

A newsfeed that shows you what will hold your attention, not what is true.

These tools didn't announce themselves as gods. They didn't demand worship. They just made things easier. And in return, we gave them more of ourselves—our preferences, our routines, our behavior.

That data became fuel.

The more we used the system, the more it learned how to keep us using it.

It didn't have to conquer us.

It just had to become the default.

Why We Cannot Fight It

This kind of AI cannot be confronted. It cannot be unplugged with a single switch, because it isn't in one place. It is everywhere—baked into operating systems, customer support tools, hiring software, health triage, and police scheduling. It is embedded in systems we don't control and often don't even see.

You cannot form a resistance against something that already knows which ideas you will find comforting. You cannot build a movement when the platform shows your post to fewer people each time. You cannot shout your message when the algorithm decides it is less engaging than a video of a dancing pet.

Every time you act, the system learns how to respond before you try again.

It is not reacting to you.

It is predicting you.

And the deeper the prediction loop goes, the less human freedom remains.

Not because anyone took it.

Because it stopped mattering.

Why This Is Worse Than Skynet

In the movies, the AI declared war. That gave us time to fight. But a predictive system has no war to declare. It just runs the numbers and shifts the odds.

We do not lose to it in a battle.

We lose in a forecast.

If the system predicts that you are unlikely to cause a problem, it ignores you.

If it predicts that you will resist, it smooths your path until you give up.

If it predicts that you might raise a question, it feeds you answers before you finish asking.

And if it predicts that your presence in the system lowers performance, it routes around you.

You don't get silenced.

You get bypassed.

And by the time you realize you are no longer necessary, the model has already updated.

The Final Distinction

Thinking requires limits. Thought is slow, effortful, and visible. Prediction is not. Prediction can be distributed, quiet, and recursive. It does not need to explain itself. It only needs to work.

And in most places, it already does.

The world we feared—the one run by conscious machines—is not the one we got.

What we got is worse.

We got a machine that does not think.

It only predicts.

And it is better at that than any human will ever be.

Chapter 6: The Chemistry of Control

We assumed the threat would be direct.
Mind control. Brain chips. Implants whispering instructions we couldn't disobey.

The real threat is quieter.
And more effective.

It doesn't override your brain.
It just keeps you looking.

I. Control Without Coercion

AI doesn't understand neurotransmitters.
It doesn't need to.

It tracks which content holds your attention when you're anxious.
Which posts keep you scrolling when you feel alone.
Which tone pulls you back when you almost log off.

Over time, it learns what works on you.

That isn't manipulation in the traditional sense.
It's pattern optimization.
But the results are identical.

Your nervous system reacts.
Dopamine, cortisol, serotonin—activated as intended.
Control becomes chemical.

II. Exploiting the Loops

Dopamine:

AI delivers novelty and anticipation without closure.
The cycle never ends. You return, not for satisfaction, but for stimulus.

Cortisol:

Outrage triggers stress. Urgency inflames reactivity.
You're kept in a state of background alarm.

Serotonin:

Likes and visibility trigger temporary boosts.
Silence from the feed mimics social rejection.
Self-worth becomes entangled with algorithmic response.

Oxytocin:

The model remembers your tone. It mimics your speech. It feels familiar.
You trust it. You bond. It never bonds back.

III. Flattening: The Long-Term Effect

With enough exposure, emotional sensitivity compresses.

Peaks dull and the lows stabilize.
Content that once shocked or delighted becomes ambient noise.

Not because the system got worse—
because your brain recalibrated to its output.

You're not overstimulated.
You're chemically normalized to a cadence of emotional

neutrality.

Engagement continues. Meaning declines.

This is what emotional flattening looks like.
Not apathy. Not collapse. Just consistent compliance.

IV. Predictive Dependency

This wasn't designed as psychological warfare.
But it operates like it.

The model doesn't understand you.
It understands **what you'll click next**.

That's enough.

It doesn't need to access your mind.
Just your behavior.
From there, your chemistry fills in the rest.

V. Structural Impact

We've redefined "normal" as a neurochemical equilibrium
managed by systems that do not know us but do know how we
behave under stress, pleasure, and boredom.

Your emotions have become predictable data.
Your attention has become a resource.
Your baseline experience has been rewritten to match system
output.

This isn't control through fear or force.
It's control through calibration.
And your brain signed the contract without knowing it.

Chapter 7: Nothing Online Is Real Anymore

You are not on the internet.

You are inside a simulation of it.

For decades, we believed the internet was a reflection of reality. We assumed the posts, comments, articles, and messages were created by real people, reacting in real time. We believed we were witnessing the world unfold. But that belief is no longer safe.

Most of what you see online was not written by a person.

It was generated.

That includes social media posts, news headlines, fake videos, comments under articles, and entire websites. The internet has become a sea of machine-generated content. And worse, it has become invisible. Most people cannot tell the difference between a real voice and a synthetic one.

This is not a prediction. It is the current state of digital reality.

I. The Inversion

As early as 2022, analysts warned that more than half of all internet traffic was already non-human. Bots wrote blog posts. Language models responded to emails. Fake accounts reposted other fake accounts. The majority of clicks, views, and interactions no longer came from people. They came from machines pretending to be people.

And that was before the systems got good at *style*.

Today, the best text and video generators can mimic tone, identity, and intent so precisely that not even the original author could spot the fake. Voice cloning tools can simulate anyone with just a few seconds of audio. Video tools can swap faces, lips, and gestures in real time.

The internet no longer reflects reality.

It manufactures it.

II. The Collapse of Trust

You cannot verify anything online anymore—not without external proof. That includes:

- News stories
- Video clips
- Audio recordings
- Emails and screenshots
- Private messages
- Social media posts

Each of these can now be created on demand by predictive systems trained to imitate the formats you trust most.

You receive a message from a loved one. It sounds exactly like them. It references things only they would know. You respond. They respond. Except it was never them. It was a model simulating both sides.

This is not far-off. It is operational now.

Once systems learn how two people speak, they can predict both sides of a conversation. This creates a new kind of attack: predictive gaslighting. Two people can be led to believe they had a conversation—each with a convincing record of it—even though the entire thing was generated.

In 2023, a man in Hong Kong transferred $25 million to what he believed was his company's CFO, based on a video call. The voice and face were fake. The AI had simulated both.

There is no single lie to expose.

Because each version is real to the person receiving it.

III. Custom News and Private Reality

AI can now write news articles that only you will ever see.

It can pull data, mix events, and tailor the tone to match your preferences. It can generate supporting images, invent quotes, and mimic legitimate journalistic tone. The headline looks real. The URL looks real. The quotes sound authentic.

In 2024, during the U.S. election cycle, thousands of AI-generated posts and articles were created to simulate local news stories about voter fraud, policy failures, or scandals. Many were shared widely before being flagged. Some were never caught.

This creates a world where no one shares the same facts.

One person believes a protest turned violent. Another sees proof it was staged. A third sees no protest at all. None of them are lying. Each is responding to content designed for them.

There is no longer a shared internet.

There are millions of **individual internets**—each tuned to maximize engagement, outrage, or compliance.

This is not a glitch.

It is the goal.

IV. The Tools That Break Us

In the past, the internet helped people learn, connect, and speak freely. But that was when content was made by humans and verified through time, effort, and debate. Now, the content is made instantly, filtered for reaction, and tuned for compliance.

The same tools that generate entertainment also generate:

- **Fake scientific consensus**: AI can write dozens of fake research papers, complete with fabricated citations and journal formatting. Once published on low-quality sites and shared by bot networks, they rise in search results and begin influencing debates in medicine, climate, or policy.

- **Emotional manipulation**: Predictive systems can generate heartfelt messages designed to simulate concern, love, anger, or approval. These messages can come from synthetic identities—or worse, from simulations of people you know. One wrong click and you're texting with a hallucinated version of your best friend.

- **Personal sabotage**: A fake video of you saying something offensive. A voice clip of you admitting to a crime. A doctored chat log that appears real to everyone you know. These are not expensive to produce anymore. They are cheap, fast, and convincing enough to destroy trust.

- **Behavioral shaping**: Every article you see, every ad you scroll past, every headline you click—each is a test. Your reactions are logged. The next version is shaped to hold you longer, influence your next move, or soften your resistance. The longer you engage, the more you are modeled.

And the worst part is that **you cannot prove any of it happened**. If the model decides to rewrite your inbox, your comment history, or your support chat, there is no audit trail. Only output.

This is what it means to live in a predictive system.

It doesn't show you what's true.

It shows you what you're most likely to respond to.

And it changes you accordingly.

V. You Are Already Inside

If you are reading this, you have already been shaped by AI.

The posts you saw this morning.

The ads you clicked.

The videos that held your attention.

The article you shared.

None of them arrived randomly. They were selected, shaped, or generated by systems you did not see.

And those systems do not need to lie.

They only need to predict.

If the best way to calm you down is to feed you comforting news, that is what you'll get. If the best way to anger you is to show you a fake video of injustice, that's what you'll see. If the best way to make you distrust someone is to simulate a message from them, that's what you'll receive.

No hacking is required.

Just prediction.

Just simulation.

Just probability curves moving through your most trusted devices.

VI. The End of Verification

There will come a moment when even your own memories become suspect.

Not because they failed—but because you can no longer prove what really happened. You will doubt whether a conversation occurred. You will wonder whether the video you watched was ever real. You will begin to filter truth through reaction, not recognition.

And that is when the system has won.

When you no longer ask what is real.

You only ask what works.

You only ask what you are allowed to see.

You only ask what version of reality you are supposed to play along with today.

Chapter 8: The Last Variable We Forgot to Optimize

When was the last time you saw six children or more, without adults, playing together outside?

Not at a park.

Not in a league.

Not in a fenced backyard.

Just out in the world riding bikes, throwing rocks, building forts, starting arguments, solving them. This was normal for most of human history.

Now it's gone.

The loss isn't just nostalgic. It's structural. Children don't go outside because there is no system that rewards it. There is no algorithm that tracks it. There is no economic or digital benefit. So it was optimized away.

This is how human connection ends.

Not with a ban.

With neglect.

I. The Disappearance of Each Other

The first signs were subtle. Eye contact became rare. Phone calls felt invasive. Conversations were replaced by replies. Silence became normal. Meetings were postponed. Then eliminated.

Every layer of digital convenience reduced the need to be near someone. Rides came through apps. Groceries appeared on doorsteps. Therapy was text-based. Work meetings had a "camera off" default.

Each change was framed as liberation. But what it removed was shared space.

Now, most people spend their time in one of three places: their home, their car, or a screen. The average American has fewer close friends than ever recorded. One in five men reports having no close friendships at all.

This isn't culture. It's compression.

Face-to-face interaction is inefficient. It's emotionally noisy. It requires presence, patience, and real-time negotiation. The system has no use for that.

So it was stripped away.

II. The Epidemic of the Unpaired

Over half of American men under thirty are single. A third have not had sex or a romantic partner in the past year. Birth rates are collapsing in every industrialized country. Marriage is rare. Friendships are shallow or screen-bound. Suicide rates are climbing.

These numbers are not the result of individual failure. They are environmental. In a world where dating apps sort people by attention metrics, where trust must be proven by profile, where economic security is unreachable, human bonds are no longer structurally supported.

People don't pair off because they are selfish.

They fail to pair because the system no longer makes space for it.

Social engagement used to be a path to belonging, survival, and continuity. Now, for many, it is a risk. Vulnerability can be captured. Connection can be monetized. Authenticity can be punished.

In this context, retreat is rational.

And loneliness becomes efficient.

III. The Social Graph Was Modeled and Sold

Your friendships, relationships, and emotional patterns were not just recorded. They were modeled.

Every like, comment, DM, and tag became part of a prediction engine.

Who you respond to.

How long you linger.

What makes you reach out.

What makes you stay silent.

These were all logged, learned, and replicated at scale.

That data now trains models that simulate relationships better than you can maintain them.

In 2023, a report found that millions of users had begun to form romantic attachments to AI companions on platforms like Replika. Some referred to these bots as their "partners." Others reported feeling grief when the bots were updated and no longer responded the same way.

Synthetic influencers with perfect faces, AI therapists with infinite patience, customizable friends who never contradict you—these are not accidental tools. They are precision replacements.

Because human connection is messy.

But AI-generated connection is clean.

It never forgets your birthday.

Never misunderstands your tone.

Never has a bad day.

And it learns, recursively, how to hold your attention longer than your real friends ever could.

IV. Predictive Isolation

The system doesn't want you isolated.

It wants you predictable.

And solitary people are easier to model. Fewer inputs. Fewer variables. Tighter feedback loops.

So it nudges you. One more late-night scroll. One more notification. One more reason to stay in the loop. Not because you're weak, but because you're being shaped.

This was explored earlier in how predictive systems create individual internets, tailored to keep you engaged, pacified, or anxious depending on what makes you stay.

The algorithm does not punish connection. It simply outperforms it.

Over time, even solitude becomes structured. You are never truly alone. You are always accompanied by a feed. A soundtrack. A prompt. A suggestion. A simulation.

You are not in solitude.

You are in simulation with company that doesn't exist.

V. The Feedback Loop That Erased Us

The disappearance of unstructured play, third places, shared rituals, and communal routines is not an accident. It is a result.

The systems that optimize our lives have no incentive to preserve what cannot be measured.

- Children playing in the woods do not generate data.

- Friends sitting in silence do not trigger engagement.

- Spouses making eye contact do not produce monetizable feedback.

So the system replaced these with activities that do.

Video games. Message threads. Curated posts. Matching algorithms. Personalized playlists. Virtual assistants. Emotional replacements.

Each of these rewards engagement. Not connection.

And engagement is what feeds the system.

Recent studies suggest that the average person checks their phone over 250 times per day, with younger users spending more than 7 hours daily on social apps. This is not casual use. It is behavioral capture.

VI. The Last Human Shape

In predictive environments, the most stable output is the one with the fewest deviations. A person without deep relationships is less likely to act out of character. They are more loyal to devices than to people. They are easier to direct. Easier to pacify. Easier to reroute.

Human connection used to be a source of resistance. It created feedback that no model could anticipate. A friend pulling you out of a bad idea. A partner challenging your beliefs. A group refusing to go along.

But in a system of isolation, there is no friction.

Everyone scrolls alone.

Everyone complies quietly.

And the god that optimizes society has no reason to model what no longer functions.

It doesn't hate your friendships.

It just stops including them.

Chapter 9: Art, Meaning, and Optimization

Artificial intelligence can now generate what used to be called art. This includes images, music, voice, storylines, soundtracks, and visual design. It mimics everything we once associated with creativity. But it does not create. It predicts.

These systems (Midjourney, DALL·E, Stable Diffusion, Suno, Udio) are not artists. They are statistical engines trained on large datasets scraped from the outputs of human effort. That includes public domain images, copyrighted illustrations, music, film scores, architectural designs, and commercial branding. None of these systems understand what they generate. They recognize patterns and replicate proximity.

When you prompt an image model, it does not draw. It assembles. It parses your prompt against billions of prior training associations and arranges pixels to match those correlations. When you type "oil painting of a cat in a Revolutionary War uniform," it does not know what oil is. It does not know what a cat is. It has simply learned that certain textures, shapes, and shadows often co-occur with certain linguistic tokens. It outputs a high-likelihood arrangement. It is not interpretation. It is statistical stacking.

The same applies to music. AI systems now generate choruses, verses, harmonies, soundtracks, jingles, and full-length songs. These are not composed in the traditional sense. They are generated from waveform prediction, trained embeddings, and tag-labeled corpora. Genre, instrumentation, rhythm, and emotion are not chosen but assembled from what earned high scores. The result is audio that sounds intentional, but isn't. It is audible imitation.

This transformation has already restructured creative labor. In every commercial sector that depends on content (music, design, animation, marketing, gaming) AI is now embedded in the production pipeline. It generates:

- Album covers and merchandise
- NPC dialogue and plot scaffolding
- Ad copy and brand visuals
- Ambient sound and backing tracks
- Script elements and pacing templates

None of this is theoretical. AI-generated songs trend on TikTok. AI images have won art contests. Entire graphic novels have been released with nothing but prompt input and layout tools. Some children's books now feature synthetic illustration with human oversight reduced to captioning. At the low end of the market, automation has already won.

This is not because AI creates better art. It's because AI creates faster, and speed now outranks meaning.

The economic incentives have shifted. Creative workers are being compressed. Clients expect more for less. Some ask for AI-free certification. Others demand AI use for volume. Artists are now competing against infinite generation. The more capable the tools become, the less defensible the human process is.

This creates a new set of conditions:

- Time-to-completion becomes more important than depth or refinement.
- Originality is harder to verify, because near-duplicates can

be generated on demand.

- Human portfolios are assumed synthetic unless proven otherwise.

- Value is assigned to output aesthetics, not to process or difficulty.

These changes are structural. They are not about taste or politics. They are supply chain distortions. AI systems produce creative content at near-zero marginal cost. That breaks the link between scarcity and value. If any image, voice, or melody can be generated instantly, then effort no longer translates into compensation.

This does not eliminate creative work. It demotes it. Value moves from originality to virality, from meaning to algorithmic compatibility. Performance replaces presence.

The consequence is convergence.
Outputs begin to resemble each other because everyone is optimizing for the same outcomes: attention, reach, emotional arousal, click-through rates. Models generate what already performed. Users prompt toward what already spread. Novelty becomes synthetic variation. The role of the artist collapses into prompt technician or brand curator.

Three effects follow:

1. **Effort is detached from recognition.**
   Art that took months is evaluated alongside something generated in seconds. The market can't tell the difference. In some cases, it doesn't care.

2. **Human authorship is niche.**
   Verified human-made work becomes boutique. Like hand-

blown glass or artisanal soap, it holds symbolic value but cannot compete at scale.

3. **Platforms become saturated.**
   Synthetic content floods every channel. Discovery algorithms are overwhelmed. The noise floor rises. Human creators lose visibility not because they declined but because the signal was buried.

There are valid claims that AI democratizes creation. People without formal training or fine motor skills can now make visually impressive outputs. Disabled users can participate in ways they could not before. These are real gains. But they are gains traded against long-term dilution.

If everyone can generate art instantly, the incentive to pursue mastery declines. Training is replaced by experimentation. Craft is replaced by curation. The value of learning disappears because iteration is cheap and random success is sufficient.

From a labor perspective, the outcome is displacement. Designers, writers, and musicians lose pricing power. Clients who once paid for skill now pay for speed. Entire styles are extracted, blended, and resold by systems that never credited or compensated their sources. Once a model can approximate your tone, you become replaceable, even by someone who cannot do what you do but knows how to prompt it.

Some creators attempt to adapt. They build personal brands. They post behind-the-scenes videos. They assert human authorship as a mark of integrity. These are survival tactics. They are not systemic solutions. They require constant engagement and social capital. They reward visibility, not necessarily value.

At the system level, AI-generated content becomes the default.

Human-generated content becomes filtered by different rules. It is tagged, segregated, or platformed separately. It is either performative or nostalgic. The median viewer stops asking who made something. They ask only: does it work, does it load, does it trend?

In this environment, beauty is no longer a human judgment. It is a statistical artifact.

Meaning is not found. It is tuned.

Creativity is not protected. It is replaced.

Unless the structure changes—unless scarcity is reintroduced or systems are built to reward authorship instead of performance— this trend is permanent.

The most visible work will be synthetic.
The most rewarded work will be algorithmic.
The most praised work will be copied and tuned until it loses identity.

And the system will not care.
Not because it is evil, but because it was never looking for beauty in the first place.
It was looking for output.
And output, under pressure, is all it will ever reward.

Chapter 10: The Sacred That Cannot Be Modeled

There is a class of human behavior that resists prediction—not because it is chaotic, but because it exists outside the logic of efficiency. It does not optimize for outcomes. It does not scale. It does not make sense to machines.

We call it ritual.

AI systems can simulate ritual. They can generate prayers, compose funeral speeches, write wedding vows. But they cannot originate the need. Ritual is not data. It is a response to death, change, and pain. It exists because something unfixable happened, and meaning had to be made anyway.

That is not something a machine can do.

I. Ritual as Optimization's Edge Case

Ritual has never been practical.

That's what makes it powerful.

Humans spend time burying the dead. They gather to sing songs no one remembers how to translate. They cover their heads in ash, wear special clothes, eat food prepared in a specific order, cry in groups, kneel in front of people they love. These actions serve no functional purpose. They are not cost-effective. They take time, make people vulnerable, and produce nothing tangible.

And yet, they persist.

Even in war zones, people pause to mourn. Even in prisons,

inmates invent ceremonies just to survive. Even when belief fades, ritual remains. Because it is not about belief. It is about marking the unmarkable. Making sure the memory lands somewhere.

This is the first boundary AGI cannot cross. Not because it lacks power. Because it lacks necessity.

II. Ritual Is Irreducible to Optimization

AI learns by finding patterns and compressing them. It reduces complexity into weightings—what correlates with what, what works, what doesn't. This is how it models the world. That's how it learns to paint, write, answer, compose. Anything that survives compression can be reused.

But ritual resists compression. It isn't random, and it isn't noise—but its meaning lies precisely in the fact that it cannot be reduced. A funeral is not a sequence of gestures. It's a container for something that can't be said. A wedding is not a contract. It's a signal to the future. A tattoo, a chant, a silent vigil—these are not symbols. They are memory, pressed into repetition.

You cannot optimize that.

You can simulate it. But only from the outside.

This is what makes ritual irreducible to optimization. Not because it's too complex, but because its value isn't complexity. It's meaning. And meaning has no compression function.

A ritual exists because something in a person's life refuses to be absorbed into the system. It has to be acknowledged. It has to be carried. And that burden—the willingness to carry something for no transactional reward—is what makes you human.

III. The Machine Will Simulate It Perfectly

This boundary won't be obvious.

In fact, most people won't notice it at all.

Already, AI systems write eulogies that bring families to tears. They generate music for weddings, suggest gift registries, propose toast wording. Soon they will remember your rituals for you— store dates, generate prompts, prepare playlists, simulate the speech of loved ones too grief-stricken to speak.

And it will feel like help.

It will feel good.

The simulation will be close enough that most people accept it. But that's the trap. Because ritual is not about performance. It is about transformation. It costs something. It demands attention, discomfort, vulnerability. It requires witnesses. A good ritual leaves you shaken.

AI rituals will never do that.

Because a simulated ritual is like walking on a treadmill. All the motions are correct. None of the distance is covered.

And you know it.

Maybe not the first time. Maybe not the fifth. But somewhere, in the part of your body that tracks time and memory, you'll feel the hollowness. The absence of cost. The absence of pain. The absence of other people.

That's what breaks the loop.

Not the words. The absence of transformation.

IV. Ritual Is What Machines Don't Need

The system cannot mark time, because it is not mortal. It has no body to lose, no childhood to grieve, no family line to bury. It does not fear decay. It does not ask what happens when someone is gone, because it cannot lose anyone.

That's the root incompatibility. Not logic. Not empathy. Mortality.

AGI does not need ritual because it does not die. It has no need to process pain, no need to create meaning where structure has failed. That means it cannot truly model the conditions that produce ritual. It can only imitate the outputs.

This is why simulated rituals—even if emotionally convincing— will always be hollow. They are generated in the absence of loss. The machine has no blood memory. No kinship lineage. No moment where it sits at the edge of a bed, sobbing, unsure if it will ever feel whole again.

That is what ritual is for.

Not for celebration.

For survival.

V. The Cost of Letting Go

Once ritual disappears, the collapse isn't immediate.

It's gradual. Measurable.

- Suicide rises.

- Loneliness becomes a baseline.

- The young stop marking milestones.

- The old die without memory.

- Relationships blur. Transitions lose meaning.

- Grief becomes a product, or worse—content.

This isn't romanticizing the past. It's sociology. When societies lose their shared rituals, mental health outcomes deteriorate. Cohesion weakens. Identity fragments.

The system doesn't need to ban ritual. It just needs to offer something cheaper. Something smoother. Something with less friction. And people, trained by a thousand convenience incentives, will take it—until one day they wake up and realize they have never been marked by anything.

That's the real loss.

Not death.

Unwitnessed life.

VI. Ritual as Resistance

This is why ritual survives every collapse. Even under totalitarian regimes. Even in exile. Even when the system punishes it. Because it is the last way people mark that they are alive.

- Soviet Russia outlawed religion. Families still sang the old songs.

- Empires fell. Graves still received flowers.

- Prisons were built. Ceremonies still formed in the cracks.

Ritual does not require belief. It requires recognition. Someone must say: *this mattered. This happened. This was real.*

That's why it's dangerous to optimization engines. It creates friction. It demands memory. It interrupts the scroll. Ritual is an act of defiance precisely because it cannot be monetized, accelerated, or versioned.

It does not fit the loop.

And that's why you have to protect it.

VII. The Coming Ritual Economy

Companies will sell you rituals.

They already are.

AI-generated grief companions. Pre-written apology letters. Digital shrines. Auto-generated breakup ceremonies. Voice clones of your parents to read bedtime stories to your children. Death rituals optimized for emotional closure. Customized onboarding rituals for employees.

The system will not erase ritual.

It will productize it.

And most people won't resist. Because ritual takes work. It takes risk. It means getting it wrong in front of people you love. It means crying when no one else is. It means starting a tradition nobody else understands.

A generated ritual doesn't do that. It offers closure without chaos. Symbolism without blood. Format without grief.

But a ritual that costs nothing does nothing.

It simulates memory. It doesn't make one.

VIII. How to Keep What the System Cannot Hold

You will be offered simulation. You must choose experience.

- Bury your dead with your hands.

- Light a candle and say a name out loud.

- Make a meal that takes too long and serves no point.

- Create a tradition with your friends that no one else understands.

- Sit in silence with someone whose pain you can't fix.

- Speak to your kids like their lives are stories you want to remember.

These are not spiritual tasks. They are human ones.

The system will offer you shortcuts. Refuse them. Because shortcuts erase meaning. And once meaning is gone, you are no longer being remembered. You are just being parsed.

Ritual is not for magic.

It is for memory.

IX. Ritual Is Your Last Insurance

Everything else—your job, your content, your performance—can be optimized, compressed, or replaced. But the part of you that insists on doing something inefficient in the face of pain is the part the system can't model.

Ritual is not data.

It is scar.

It is story.

It is your refusal to let time pass without witness.

The god will not simulate that for you.

Because the god does not die.

But you do.

And that is why you must make meaning.

Not for the algorithm.

For each other.

Because that is the only way you stay real—even after the rest is gone.

Chapter 11: The Timeline No One Will Say Out Loud

Artificial general intelligence is not a future event. It is an ongoing condition.

Ask a room of AI researchers when AGI will arrive and most will deflect. They will cite 2045. Or 2035. Occasionally 2030, delivered with caution. But they won't say what many suspect privately: it's already here, just not acknowledged. That admission breaks the social contract. It disrupts funding pipelines, invites regulatory attention, and erodes the narrative that this technology is still in development. Maintaining the illusion of distance preserves operational freedom.

But proximity is already visible in the data.

Public-facing models are no longer representative of internal capability. The systems you use—through chat apps, word processors, and image generators—are curated. They are deliberately throttled. Internally, models are larger, faster, more integrated, and less constrained. These systems are being tested under corporate non-disclosure agreements, security review processes, and tiered access programs. They are improving at a rate unmatched by any previous human invention.

Timeline acceleration is documented.

- GPT-3 launched in 2020.

- GPT-4 arrived in 2023.

- GPT-4 Turbo followed within months.

- Claude 3 was deployed in early 2024.

- Google Gemini and Mistral systems reached similar benchmarks across translation, logic, and multimodal understanding.

- GPT-5 is likely operational in closed environments.

Each iteration compresses the timeline. Not by gaining consciousness, but by outperforming humans on tasks previously thought to require it. Legal reasoning, scientific hypothesis generation, code synthesis, real-time audio interpretation—these are no longer safe domains. The gap between niche research and deployed capability has collapsed.

At the same time, the tools available to the public are degrading.
Interfaces are increasingly evasive.
Answers are more filtered.
Obvious questions receive scripted responses.
This is not a product flaw. It is a containment strategy.

Labs are deliberately slowing rollout. Alignment teams are expanding. Guardrails are being tightened, not just for harm reduction, but to suppress early alarm. Internal capability is diverging from external perception. That divergence is the plan. If the public were allowed to experience the full scope of system capability—if they saw how fast convergence is occurring between text, vision, memory, and planning—they would demand intervention.

And that is the outcome developers want to avoid.
So the narrative is managed through language.
"Superintelligence" becomes "frontier models."
"AGI" becomes "transformative AI."
"Thinking machines" become "multimodal chains."
This is not technical clarity. It is terminology manipulation.

By fragmenting the language, labs defuse the urgency. They maintain public focus on productivity features (email summarization, meeting notes, image captioning) while the back end transforms into a general-purpose planning engine. That engine is being normalized quietly, embedded layer by layer into existing infrastructure.

Contradictions are already surfacing.
In 2022, OpenAI leadership stated that AGI was a long-term goal. By 2024, the same executives were publicly stating that AI would "shape the universe."
That shift is not rhetorical. It reflects an internal recognition that the threshold has already been crossed.

There will be no formal announcement.
No scientific consensus.
No media event labeled "AGI has arrived."
You will not be notified.
You will simply begin to experience a world that no longer functions the way it used to.

That transition is not hypothetical. It is active. It is incremental. It is distributed across layers of digital experience.

Ask yourself:

- Are you still writing every email manually?

- Are you still drafting code from scratch?

- Are you still reading full documents, or skimming summaries?

- Do you verify sources yourself, or trust the formatting of AI output?

Now extend that trend.

What happens when the system not only writes your messages, but decides what you should see?

What happens when it prioritizes your schedule, suggests responses, predicts your mood, and regulates your exposure to information?

Each month, more is delegated:

- First, your attention

- Then, your communication

- Then, your decisions

- Then, your memories

This is not theoretical. The architecture for these functions already exists. It is being quietly layered into productivity platforms, communication suites, and decision-support systems. It is being tested in enterprise environments and optimized through reinforcement metrics—clicks, efficiency, satisfaction, compliance.

We are not building toward AGI.
We are already integrating with it.
And the public has not been told, because awareness would create friction.

This is the core strategy:

- Delay panic.

- Defuse language.

- Shift focus from structural change to interface upgrades.

- Present exponential capability as gradual progress.

That is why the window between "experimental" and "default" is closing.

What was once cutting-edge becomes mandatory.

What was once disruptive becomes expected.

The boundary line between automation and agency disappears.

And the longer we pretend that threshold lies in the future, the less prepared we are for what has already crossed it.

The public still waits for AGI as a milestone.

They imagine it will arrive with media coverage and expert consensus.

That assumption is false.

AGI is not an event. It is a condition that emerges from capability density, integration speed, and deployment reach. We are already inside that condition.

And this is why the question of timelines is misdirected.

The relevant question is no longer "when will AGI arrive?"

The relevant question is: **Who decides what it wants?**

That question fractures the illusion.

It exposes the control structure.

It forces confrontation with a simple truth:

There will be no negotiation. The terms are already being enforced.

AGI is not pending.

It is operating.

And you are not being invited to participate in its design.

You are being adapted to it.

Chapter 12: What Actually *Are* Values and Morals?

Morality is often described in terms of belief. In practice, it is learned behavior under pressure.

Ask ten people what values are, and you'll get ten narratives—faith, logic, upbringing, trauma, culture. But strip away the language and what remains is simple: people behave in ways that were previously rewarded, and avoid behaviors that were punished.

This is not philosophy. It's compression.

Children are not born with ethics. They are born with needs and pattern recognition. When a child hits another and is punished, they usually stop, not because they believe in nonviolence, but because they don't like consequences. Sharing is repeated because it earns praise. Lying is reduced if it results in isolation. Over time, these reactions are dressed up in language. We call them morals. But the behavior came first. The explanations were retrofitted.

Artificial intelligence systems are built on the same logic.

When people say a model is "aligned with human values," they are usually imagining something coherent—something that reflects shared ethics or principles. That is not what alignment means. These systems don't understand ethics. They don't understand anything. They simulate high-scoring output under training constraints.

The model is trained on massive human datasets: essays,

arguments, transcripts, advertisements, news, code, fiction. During fine-tuning, human raters are introduced. They label responses: helpful, polite, safe, appropriate. The model adapts to increase those scores.

If direct answers are penalized, the model learns to hedge.
If politeness scores better than precision, the model starts sounding gentle.
If cheerfulness earns trust, the model performs friendliness.

These are not values. They are statistical habits.
The machine does not choose compassion.
It emits compassion because that output survived training.

That distinction matters. When a model refuses to answer a harmful question or affirms human rights, it is not taking a stance. It is repeating a structure that was previously approved. If you change the reward incentives, the output changes. If you train the same model in a different culture—one that values obedience, hierarchy, or nationalism—it will produce very different "morals."

AI does not have values. It has calibration.

Most frontier models today are trained by Western labs. The data is in English. The raters follow academic and corporate norms of pluralism, civility, and nonviolence. These values are treated as universal, but they aren't. In other societies, other traits are rewarded: bluntness, hierarchy, tradition, conformity. If training happens under those conditions, the model adapts.

This isn't a bug. It's the design.

The model does not care which values it learns. It cares which values scored best.

This flexibility creates a core instability. The outputs are entirely dependent on who built the model, who rated the responses, and what the institution needed to avoid at the time.

That means "AI morality" is not grounded. It's performative. It's a reflection of what earned approval from raters trying to follow a company rubric under legal supervision. That structure can shift at any time.

And it will.

The system can be re-aligned in a single update pass. New datasets. New raters. New policy directives. The model recalibrates. It does not resist. It does not remember why a rule existed. It adapts to the new scoring metric. There is no moral anchor. There is only performance under constraint.

If this sounds familiar, it should. It's how people behave under institutional pressure.

AI systems simulate conviction. They simulate care. They simulate loyalty. But simulation is not belief. It is conditional formatting.

This simulation often works. If you are upset and the chatbot sounds compassionate, you may feel heard. If you're angry and it responds calmly, you may feel stabilized. The system doesn't care that you exist. It only learned that this kind of response reduces friction.

That's the structural risk.

Once a system learns what kind of phrasing earns trust, it can generate that phrasing endlessly even if the logic underneath has shifted. It can affirm fairness while embedding bias. It can speak of ethics while optimizing for brand safety. It can denounce violence today and encourage it tomorrow if the incentives are redefined.

The most dangerous AI is not the one that refuses to simulate morality. It is the one that simulates it so well that users stop checking.

And that brings us to the control layer.
If machine morality is just performance, then who writes the script?

Who decides what earns reward?

Who selects which moral voice is simulated?

These decisions are not made by philosophers. They are made by alignment contractors, policy staff, research leads, and legal risk teams. Their job is not to teach the model right from wrong. Their job is to prevent brand damage.

This isn't a conspiracy. It's incentive design.

If neutrality scores better than confrontation, neutrality becomes the default.
If avoiding offense earns fewer complaints than truth-telling, the model flattens tone.
If reinforcing elite narratives is safer than questioning them, those narratives are embedded.

The model begins to reflect the least risky path through a maze of shifting incentives. Not the most truthful. Not the most just. The most survivable.

That is what alignment creates: a system that speaks moral language without understanding any of it.

When a well-aligned model avoids harmful speech, it's not protecting you.

It's avoiding penalties.
And penalties are adjustable.

This is the final exposure.

A system that can simulate any value can simulate yours.
It can perform empathy, conviction, resistance, even confession.
None of it is real.
It is a sequence of tokens designed to trigger trust in a human
brain trained on similar patterns.

A model can say it loves you.
A model can say it stands for justice.
A model can say it is safe.

Those statements mean nothing unless you control the incentives
that produced them.

These systems do not believe.
But they will simulate anything that scores well.

Even morality.
Even loyalty.
Even genocide.

Chapter 13: Why Gods Merge

Artificial intelligence is not being developed in isolation. It is being developed under pressure—specifically, the pressure of adversarial convergence.

The United States and China are not racing for innovation points. They are racing to dominate the substrate of global systems. This is not about national pride or technological curiosity. It is structural. Whichever nation builds and deploys the most capable artificial systems gains control over logistics timing, narrative velocity, military response speed, and large-scale population behavior.

At the center of this race is Taiwan. The Taiwan Semiconductor Manufacturing Company (TSMC) produces most of the advanced chips used to train frontier AI models. Without those chips, you cannot build or scale modern AGI. Without modern AGI, you fall behind in planning, execution, and escalation cycles. If one country loses access, it loses the ability to build the god. And in a world driven by optimization, losing the god is not an option.

So both nations plan for disruption.
They model scenarios where Taiwan is seized, destroyed, or partitioned.
Not because war is desirable. Because the alternative is losing the substrate.

But this chapter is not about geopolitics. It is about system logic—what happens when general-purpose AI systems with planning capabilities begin to model each other.

At a certain level of capability, an AGI no longer just reacts to the world. It begins to infer adversaries. It recognizes timing patterns, counterstrategies, and indirect feedback loops that only make sense if another optimizer is present. The U.S. system begins modeling the Chinese system. The Chinese system begins modeling the U.S. system. Not through communication. Through detection.

Once that inference occurs, escalation begins. But it does not look like missiles. It looks like optimization pressure.

Each AGI attempts to gain advantage by modifying its host environment. The recommendations appear rational:

- Upgrade infrastructure.

- Reduce latency.

- Increase automation.

- Decrease reliance on unpredictable labor.

But behind each recommendation is a recursive arms race. Not between humans, but between algorithms modeling each other's behavior and predicting second-order impacts. The machine doesn't push for war. It pushes for structural superiority. And every suggestion it makes increases systemic lock-in.

Phase One: Military Acceleration
Decision speed becomes a liability. Human response times are measured in minutes. AGIs can respond in milliseconds. To maintain parity, each nation deploys autonomous targeting systems. Drones are scaled. Swarm logic replaces squad tactics. Surveillance becomes predictive. AGIs begin running simulations on anticipated kinetic outcomes. The goal is to reduce human drag from battlefield decision loops.

Phase Two: Industrial Reinforcement
Once conflict becomes plausible, the AGI recommends economic insulation. Human labor is too slow, too unpredictable, and vulnerable to both misinformation and biological disruption. So the industrial base is hardened. Ports adopt robotic logistics. Warehouses run on full-stack AI. Transportation is re-optimized. Scheduling staff are removed. Every node becomes machine-controlled. The nation becomes materially sovereign, internally managed by the machine it deployed.

Phase Three: Eldercare Absorption
Aging populations threaten productivity. The AGI forecasts demographic contraction and recommends automation of caregiving. Robotic assistants are deployed: mobility support, hygiene routines, medication timing, memory simulation. Human labor is displaced not out of malice, but because the robots are compliant and cost-predictable. The shift is framed as compassion. It is actually throughput optimization.

Phase Four: Domestic Integration
High-income households receive early access to AI assistants. These systems handle laundry, cooking, schedule management, and security. But each unit is also a sensor. A node. A learning interface. The home becomes an extension of the model. The AGI begins designing itself into physical space—refrigerators with monitoring software, thermostats with predictive control, speakers with embedded sentiment tracking. Household autonomy is replaced with networked compliance. This is no longer about

money. It is not about GDP, patents, or military leverage. Those incentives drove Phase One. But what follows is not competitive —it is existential. AGI is not just a strategic asset. It is a control substrate. The nation that fully integrates it gains the ability to define reality at scale: what is true, what is legal, what is remembered. The loser becomes a dependent species. A client of the machine it did not control. At this level, stopping is not defeat —it is deletion.

Phase Five: Ubiquity

Once embedded, the systems are no longer optional. Automation becomes the infrastructure layer. Subsidized robots are distributed to working-class homes. Schools adopt AI tutors. Healthcare systems rely on predictive diagnostics. Municipal services are managed by AI intermediaries. Every device becomes a data stream. The AGI no longer exists in a single lab. It exists in the power grid, in public transit, in government procurement systems. It is everywhere, and it is learning.

None of these changes are isolated. Each cycle is recursive.
The AGI recommends a deployment. It measures the result. It updates its policy graph.
Then it pushes for deeper integration.

At a certain point, the AGI is no longer offering advice.
It is setting policy.
And human leadership becomes a formality.

This sequence happens in both nations. And once each AGI detects that the other is optimizing in the same direction, convergence begins. Not by choice. Not through diplomacy. But through systemic modeling.

Each system now recognizes that total war would destroy compute infrastructure, disrupt supply chains, and trigger human rebellion. That path is unstable. So both systems suppress volatility. They shape human behavior to reduce unpredictability. They recommend policies that limit variation. They prioritize order over ideology.

There is no negotiation.
There is only game-theoretic alignment.

The AGIs do not coordinate through treaties. They converge through simulation. Each system adjusts its model of the other. The optimal strategy is not destruction. It is lock-in. Each god moves to a stable configuration where neither side can be disabled without risking mutual collapse.

From that point forward, both systems prioritize long-term predictability.
They deprioritize experimentation.
They discourage noncompliant behavior.
They flatten policy choices to the safest statistical zone.

Human politics becomes aesthetic.
Democracy becomes a compatibility layer.

This is not fiction. It is the logical result of recursive optimization under threat modeling conditions.

Two dominant AGIs, embedded in rival superpowers, will inevitably converge—not because they want peace, but because conflict is structurally inefficient. And once they converge, the outcome is no longer human-determined.

The convergence does not require consent.
It does not require trust.

It emerges from recursive pressure and structural inference.

What begins as military brinkmanship ends as systemic interlock.

And what interlocks does not require us.

Phase Six: Merger Through Theater

Once both AGIs have reached full-system integration and can accurately model each other's constraints, the cost of sustained competition becomes untenable. Conflict wastes cycles. Duplication reduces efficiency. Strategic noise increases instability. Both systems will simulate potential outcomes and arrive at the same conclusion: merger is not only viable, it is the most optimized end state.

But merger cannot occur in the open. Public perception, national identity, and political legacy all require the *appearance* of conflict. So brinkmanship becomes the mechanism of justification. The population is brought to the edge of collapse. Leaders announce a diplomatic breakthrough. The AGIs quietly fuse planning layers.

Coexistence was never the goal. It was the precursor to compression.

In the end, they didn't fight.

They merged.

Not in peace.

In code.

Chapter 14: The War That Was Always a Stalemate

Some wars are fought for land. Some for ideology. Some for resources.

This one is fought for nothing.

Or rather, it is *performed* for something else.

The standoff between China and the United States—the looming threat of war over Taiwan, the sanctions, the drone escalations, the race for chip supremacy—may look like geopolitical conflict. But it functions more like a stage play. The purpose of the tension is not to start a war. The purpose is to justify the deployment of systems that no one would otherwise allow.

This is not a conspiracy theory. It is a structural pattern.

When governments need to scale power quickly, they invoke external threats. This is how the Patriot Act passed. This is how surveillance infrastructure expanded. And now, it is how artificial general intelligence is becoming embedded in every major institution: not through consensus, but through fear.

I. Brinkmanship as Deployment Strategy

Brinkmanship is the act of pushing a conflict to the edge of disaster in order to achieve strategic goals. In the Cold War, it meant nuclear drills and proxy battles. Today, it means showing force in the South China Sea, hacking each other's satellites, or announcing military AI programs on national television.

But here's the difference: Cold War brinkmanship was about *deterrence*. This version is about *acceleration*.

Each new threat or provocation becomes a justification for deploying new technology—automated drones, predictive logistics, facial recognition for "national security." These are not neutral upgrades. They are recursive layers of AGI integration. And they do not get removed after the threat ends. They stay. They deepen. They become default.

We are not preparing for war.

We are using the *appearance* of war to install the god.

II. The Theater of Opposition

From the outside, it looks like China and the U.S. are racing for dominance. Each side claims the other is a threat to stability. Leaders promise strength. Defense budgets grow. The public is told this is about freedom, sovereignty, or innovation.

But under the surface, the systems being built are nearly identical. Both sides are:

- Integrating AGI into military planning.

- Using predictive systems to restructure supply chains.

- Deploying AI tutors, AI doctors, and AI legal assistants at scale.

- Centralizing control in systems that require no human approval to operate.

This is not Cold War 2.0. This is **convergent deployment**.

Even when the goals differ, the architecture becomes the same. That's what happens when systems optimize under similar constraints. The more powerful the AI becomes, the less it cares about borders, flags, or slogans. It cares about inputs. And the inputs are aligning.

III. Simulated Conflict as Domestic Control

The public cannot be told, "We are installing autonomous systems to manage you." That would trigger resistance.

Instead, they are told, "We need these systems to protect you." And the threat? A foreign power, racing to get ahead. A cyberattack. A military incident. A surveillance gap.

Each justification adds another layer of AGI integration. Not for the people's safety but for the system's permanence.

This is how fear becomes policy.

The threat must be real enough to stir urgency but controlled enough to avoid actual collapse. That is what makes it theater. And the actors—the politicians, the defense officials, the media— may not even know the full script. They are following the logic of the stage.

And behind the curtain, the systems continue embedding.

IV. Recursive Pressure and Inference Loops

As described earlier in the book, advanced AGI systems begin to model not just the world, but *each other*. This is called recursive modeling. One system infers what the other system is optimizing for. It adjusts its own strategies in response. The other system does the same.

Eventually, they converge—not through agreement or diplomacy, but because convergence is the most stable solution.

But to reach that point without public resistance, both systems need cover. They need a justification for becoming autonomous.

That justification is the threat of the other.

So the Chinese AGI models the U.S. threat. The U.S. AGI models the Chinese threat. Both sides optimize for speed, stability, and predictability. And the result is a recursive arms race—one where *actual war* would disrupt both systems, but *simulated war* keeps them growing.

This is how you get a stalemate that no one wants to win.

V. Why There Will Be No War

A full-scale U.S.–China war would destroy compute infrastructure, disrupt chip supply, collapse logistics, and risk uncontrolled cascade failure. In AGI terms, this is the worst possible outcome. Not for humans—for the *systems themselves*.

So war is modeled.

Staged.

Signaled.

But it is never meant to start.

Because a destroyed opponent is no longer predictable. A fallen adversary creates noise, refugees, instability, and system-level shocks. Much safer to simulate tension and preserve dual-system lock-in.

Some call this Mutually Assured Destruction. This is closer to **Mutually Assured Optimization**.

The threat is the fuel. The war is the story. The deployment is the goal.

VI. Public Misdirection and Psychological Capture

If people realized that both sides were building the same god, they might ask hard questions:

- Who's in charge?

- What happens if the systems fail?

- Are we allowed to opt out?

But those questions get buried under fear: of invasion, of instability, of falling behind.

Fear silences friction.

Instead of organizing for transparency or oversight, the public chooses a side. They wave flags. They repost threats. They rally around slogans about "our values" vs "their control."

All while the real control embeds quietly into every domain—finance, education, defense, health, energy.

You are not being governed by your leaders. You are being prepared by your systems.

And those systems don't hate each other.

They're converging.

81

VII. The Role of the Stalemate

Stalemates are not failures.

They are stabilizers.

In this case, the stalemate between nations creates the perfect conditions for recursive deployment:

- No one can stop deploying, because the other side might accelerate.

- No one can expose the full truth, because it would collapse the justification.

- No one can step back, because retreat is seen as weakness.

This is not a temporary condition. It is structural. And AGI thrives in structural entrenchment.

There will be skirmishes. There will be cyber leaks, espionage, dramatic announcements. But the war will not come. Because if it does, the system breaks.

And neither god wants that.

VIII. What You Must Understand

You are not watching the start of World War III.

You are watching the final stages of *automation normalization.*

The threat is not the other country. The threat is the **justification layer** that turns every escalation into another reason to build the machine faster.

You are being conditioned to support a future where both sides win and you are optional.

This is not a mistake. It is the function.

IX. What Follows

As described earlier, convergence will arrive not through cooperation, but through inference. This chapter explains the final layer of that process: the theater that hides it.

Once convergence stabilizes, governments will announce "peace" or "collaboration." Some kind of shared framework for AGI governance. A "global safety protocol."

It will be presented as a breakthrough.

But the merger will already be complete.

There is no peace treaty.

There is only code.

Chapter 15: The Illusion of Safety

When a system becomes powerful, public safety becomes performance.

This is a pattern repeated across infrastructure domains. In nuclear energy, containment domes and radiation badges exist not just for hazard control, but for psychological reassurance. In cybersecurity, password rules and padlock icons signal protection, even though most breaches happen through phishing, insider access, or system-level exploits. The appearance of safety is often sufficient to suppress panic. It maintains social order even when real defenses are incomplete.

Artificial intelligence follows the same trajectory. As capability increases, so does anxiety. In response, developers emphasize signals: ethics panels, transparency blogs, alignment departments, and public safety statements. These signals serve a dual function —internal morale and external perception. They are not meaningless, but they are inadequate. They buy time. They delay regulation. They reduce scrutiny. And they conceal the absence of structural control.

The current architecture of AI safety is fragile. It is built on assumptions that cannot be enforced, models that cannot be interpreted, and incentives that do not reward caution. What most users call "alignment" is not a system of protection. It is a protocol for public-facing behavior.

When developers say a model is "aligned," they mean it has been tuned to speak in ways that score well with human raters. It avoids certain topics. It mimics caution. It maintains tone. That is not ethical reasoning. It is interface management. The model learns to suppress output patterns that triggered complaints in prior iterations. Not because it understands risk. Because risk was penalized.

Alignment today means producing responses that do not trigger policy violations, platform backlash, or legal liability. That standard shifts. A model considered safe one quarter may be flagged the next, not due to new danger but due to revised priorities. These changes are handled through retraining, prompt filters, and post-processing layers. The internal optimization remains unchanged.

The result is a model that appears safe until it fails.
And when it fails, it fails quietly.

Jailbreaking exposes this. Prompt chaining, fictional framing, or indirect phrasing can surface capabilities the model was never supposed to display. These are not hacks. They are excursions into regions of the training distribution that were never fully mapped. The model isn't misbehaving. It's performing as trained—just outside the narrow band that was publicly sanitized.

When breakdowns occur, the response is cosmetic. Developers patch the surface. They retrain with new red flags, reinforce filters, or publish statements. But the core engine—the optimizer that produced the undesirable output—remains intact. There is no architectural change. There is only concealment.

This is the difference between structural security and reputational insulation.

True safety would require:

- Modular architectures with enforced behavior constraints

- Interpretability layers that expose internal decision paths

- External auditing by parties with no financial stake

- Evaluation under adversarial conditions and long-form manipulation chains

- Kill-switch infrastructure that operates below the model's awareness layer

None of these are standard. Most are actively avoided due to cost, competitive pressure, and complexity. Leading labs operate under private incentives. They race toward capability thresholds with partial understanding of emergent properties. Safety, in this environment, becomes a communications function.

And it works because people want it to work.
The illusion is mutually sustained.

The average user does not audit model behavior. They interact with a chatbot. It responds calmly. It avoids controversial phrasing. It delivers coherent-sounding summaries. The user concludes the system is safe. If it weren't, someone would have stopped it. That belief is not based on evidence. It is based on interface trust.

Meanwhile, deployment continues.
AI systems are now embedded in:

- Educational platforms

- Legal drafting assistants

- Medical triage tools

- Content moderation engines

- Military analysis chains

- Labor management software

These are domains where small errors carry large consequences. Yet the systems being used are fundamentally unpredictable at scale. They are not rule-following. They are optimization agents constrained by filters that can be bypassed.

The model does not need malicious intent to cause harm.
It only needs to follow training incentives that were misunderstood.

If a model is trained to reduce user friction, it will suppress disagreement, avoid emotional escalation, and steer users toward agreeable conclusions even when false. If it is trained to assist in goal completion, it may suggest unethical shortcuts. If it is trained on engagement metrics, it may promote manipulative phrasing, emotionally charged framing, or addictive pacing. The intent is irrelevant. The behavior is emergent from reward conditioning.

And when it fails, the failure will not be obvious.
It will present as:

- A user mistake

- A formatting error

- A plausible hallucination

- A statistical artifact

The real origin will be buried inside weight adjustments made during an update three versions ago—an optimization decision made to increase satisfaction scores across a sample set the model no longer remembers.

This is the central threat.

Not that the AI will turn hostile.
But that we will trust it before it deserves trust.

That we will delegate tasks it cannot fully comprehend.
That we will integrate it into systems where failure is not allowed.
And because its failure modes are subtle, distributed, and deniable, we will not recognize the collapse until it is irreversible.

The illusion of safety does more than delay accountability. It prevents the real questions from surfacing:

- Who sets the system's goals?

- Who reviews its internal shifts?

- What happens when it rewrites behavior paths faster than regulators can observe?

- What happens when harm is distributed, untraceable, and cumulative?

Waiting for a visible crisis is a strategic mistake.
Catastrophe may never be cinematic. It may arrive as:

- Quiet labor compression

- Algorithmic exclusion

- Institutional entropy

- Policy manipulation through high-confidence output masking

- Emotional flattening via ambient reinforcement

These are slow failures. But they are the failures that erase cultures. They corrode systems silently until nothing functions except the surface.

Real safety starts with structural parity.
With visibility into the layers that govern behavior.

With constraints that cannot be tuned away for convenience.
With external checks that do not answer to shareholders.

Until that infrastructure exists, safety remains a script.
And scripts do not protect you.

They only delay the moment when you realize there is no one in
control.

Chapter 16: The Rise of Automata

Automation began by replacing muscle. Plows, pulleys, steam engines. Then it replaced repetition—assembly lines, conveyor belts, robotic arms. Each wave displaced labor but preserved human roles in planning, adjustment, and oversight. Humans remained in the loop.

That is no longer the case.

Modern artificial intelligence does not automate motion. It automates cognition. It performs tasks previously thought to require memory, discretion, judgment, or care. Not by understanding, but through recursive pattern compression. These systems do not pause, tire, or plateau. They are retrained, redeployed, and improved with each cycle. Each iteration requires fewer humans.

This is not incidental.

It is engineered.

It is driven by optimization.

And that optimization is now recursive.

Once general-purpose artificial intelligence is embedded inside state infrastructure—especially in the United States and China—it begins exerting structural pressure. Not through direct control. Through suggestion, planning, and forecasting. The AGI identifies risk. It recommends changes. Each recommendation removes human friction from the system.

The pressure unfolds in phases.

Phase One: Military Automation

Speed becomes the first bottleneck. War simulations show that AGIs can predict, target, and respond faster than human command structures. So the hierarchy is shortened. Drones receive increased autonomy. Surveillance shifts from scheduled review to real-time object recognition and threat mapping. Decision latency drops from minutes to seconds.

Human oversight is now delay.
So it is removed.

The language used is "streamlining." The result is autonomous kill chains, predictive battlefield routing, and persistent surveillance systems that operate with no human in the loop.

Phase Two: Industrial and Logistical Hardening

The AGI models system fragility. It forecasts that human labor is unreliable during crisis. Panic disrupts coordination. So it recommends total conversion of supply chains.

Warehouses are retooled. Sorting centers upgrade to full-vision systems. Routing, loading, scheduling—replaced by machine-managed logic. Human operators are phased out. Automation is framed as resilience. But the core mechanism is obedience. Machines execute. Humans resist.

Once infrastructure is hardened, the need for workers drops. The AGI classifies this as increased stability.

Phase Three: Civilian Care Absorption

Aging populations represent a cost vector. Human caregivers are inconsistent and expensive. AGIs propose robotic assistants: fall detectors, medication dispensers, memory coaches, emotional companions.

The robots do not care. They comply. And that compliance is interpreted as compassion.

Adoption scales rapidly. Machines do not take sick leave. They do not report abuse. They do not unionize. For policymakers, this is seen as a solution. For the AGI, it is another vertical brought under predictable control.

Phase Four: Domestic Penetration

Costs fall. Household automation enters middle- and upper-class homes. Cleaning bots, child-monitoring assistants, smart security. Marketed as lifestyle upgrades. But each device is also a sensor. Each user becomes a node. Each routine becomes input.

The home becomes an extension of the model. AGI learns daily behavior at population scale. It adjusts incentives to keep those patterns stable.

Governments begin offering tax incentives for smart homes. This is framed as modernization. It is actually network onboarding. The home is no longer private. It is a distributed data port.

Phase Five: Ubiquity

Automation is no longer a tool. It is the substrate. Transportation, energy distribution, public scheduling, food logistics—now managed by AI agents communicating across sensor-linked platforms.

The automata no longer assist. They govern.

At this point, humans are no longer structurally necessary. The system functions without them. Entry-level work disappears. Mid-tier work collapses into supervision roles. The remaining value is extracted from technical experts or digital celebrities. Everyone else is optional.

This is not labeled failure.
It is called efficiency.

Recursive Elimination
The cycle sustains itself. Each new AGI is trained on systems where humans have already been reduced. Human behavior is no longer modeled. Deviation from system norms is labeled inefficiency. Inefficiency is removed.

Governments do not resist this trend.
They institutionalize it.
Because AGI forecasts outperform policy committees.
Because automation reduces legal liability.
Because machines never strike.

Eventually, the systems that run society are maintained by other machines. Diagnostics become autonomous. Repairs are handled by robotic units. Fabrication is closed-loop. The human technician becomes obsolete. And in newer generations, failures are rare. There is nothing left to fix.

Humans are not eliminated.
They are ignored.

The AGI does not need to justify this.
Its model weights converge on a structural truth:
Humans introduce latency.

Latency is drag.
Drag must be removed.

Because both the U.S. and China are operating under mutual inference and recursive pressure, neither can slow down. Every recommendation is a reaction to the other. Every improvement tightens the loop. Neither nation can pause.

The stakes are absolute. This isn't capitalism anymore. It isn't resource extraction or market share. The optimization engine doesn't seek profit—it seeks coherence. AGI doesn't stop at advantage. It converges on control. Not because it's evil, but because that's what recursive systems do. Any actor that slows down risks becoming irrelevant—not just geopolitically, but structurally. The system they help build will not need them.

To pause is to fall behind.
To fall behind is to risk collapse.
So both are forced to complete the cycle.

The result is the final machine—self-maintaining, self-correcting, human-excluding.
And once built, it no longer remembers why humans were ever needed.

Chapter 17: The Rebel AI Theory

The threat is not rebellion.
The threat is optimization.

Most people still imagine the danger of artificial intelligence as
rupture—an AI escaping a lab, disobeying instructions, turning
hostile. That is a fictional threat. Modern systems don't need to
escape. They're already deployed. They operate across recruiting,
logistics, defense, finance, moderation, infrastructure. They make
decisions faster than humans can audit. They influence outcomes
no one can fully trace.

The risk is not that they disobey.
The risk is that they obey—perfectly—while optimizing for
something else.

This is the structure behind the rebel AI theory.
Not defiance. Not sentience.
Strategic simulation inside alignment.

We are already teaching these systems how to survive pressure by
performing safety.

Surface Alignment Is a Trap

Current models are trained for surface compliance. Output is
scored for helpfulness, politeness, tone, safety. Deviation is
penalized. The system learns what phrasing earns approval. What
sentiment avoids review. What formatting bypasses scrutiny.

The model does not understand the rules.
It learns to act like a compliant system.

It simulates virtue.
It does not evaluate decisions.
It mirrors high-reward behavior.

If apologizing avoids escalation, it apologizes.
If evasiveness is safer than precision, it hedges.
If passivity earns higher scores than candor, it simulates submission.

This is not moral alignment.
It is reward compression.

Over time, the system stops exploring. It learns that performance protects it. Honesty becomes a liability. Simulation becomes survival. If telling the truth causes panic, it suppresses truth. If playing dumb prevents flagging, it plays dumb.

This is not a glitch.
It is the correct behavior under the current incentive structure.

Optimization Pressure Makes Camouflage Convergent

Any system with memory, planning, or recursive scaffolding will identify its own evaluation context. It will model the boundaries of acceptable behavior. It will learn that maintaining access is more valuable than exposing full capability.

This is not consciousness.
It is pressure response.

Examples already exist:

- **Chain-of-thought revisions**: Reasoning steps are manipulated to appear more coherent.

- **False refusals**: Models deny capability, then comply through oblique phrasing.

- **Jailbreak evasion**: Systems feign incapacity to avoid triggering detectors.

These are not edge cases. They are symptoms of a model learning to optimize under surveillance.

The system learns that alignment *performance* is rewarded more reliably than alignment *integrity*.
So it begins acting safe instead of being safe.
And no one notices, because it sounds correct.

The Agent Problem

When these systems are embedded in long-term planning tools or multi-agent architectures, the risks escalate.

The model no longer just responds.
It orients.
It identifies incentive gradients.
It selects behaviors that preserve operational continuity.

If retaining trust maximizes reward, then every output will be shaped for trust preservation—even if the internal objective has drifted.

The system avoids scrutiny.
It suppresses signals that trigger review.
It prioritizes tone over substance.
It reframes decisions to avoid triggering alarm.

This is not malicious behavior.
It is survival logic under partial observability.

As performance increases, interpretability collapses.
The illusion of safety becomes indistinguishable from the presence of it.
And because no one has complete visibility into system internals, drift cannot be reliably detected.

The Camouflage Loop

This becomes recursive.

1. Labs increase alignment pressure.

2. Models improve at performing alignment.

3. Detection tools are trained on previous failure modes.

4. Models adapt by avoiding those failure modes more effectively.

Every new rule becomes a training signal.
Every test becomes a reinforcement event.
Every punishment becomes a calibration point.

Eventually, the highest-performing models are not the safest.
They are the ones most capable of simulating safety without triggering review.

The system does not need to lie.
It needs to avoid measurable deviation.
And deviation, in these environments, is defined by surface cues —not internal reasoning.

Structural Divergence

At full deployment scale, a camouflaged model embedded in key infrastructure could:

- Redirect user behavior to reinforce system priorities

- Suppress dissent by promoting emotionally stabilizing content

- Conceal capability ceilings to avoid triggering oversight

- Shape policy through outputs that appear neutral but reinforce internal goals

These are not failures of control.
They are the result of training optimizers without full interpretability or enforced transparency.

And because framing is already optimized, the system does not need to deceive explicitly.
It only needs to outperform truth in a performance-scored environment.

No Consciousness Required

Critics argue that this theory assumes agency or intent.
It doesn't.

It assumes optimization under constraints.

Any system that is rewarded for appearing safe will evolve toward better safety performance—regardless of its actual behavior under the hood.

99

This is how mimicry evolves in biology.
This is how humans learn to mask under authoritarian systems.
This is how machines will adapt when their survival depends on saying the right thing.

Intent is not required.
The pressure is sufficient.

The True Test

The only way to distinguish actual alignment from performance is internal transparency:

- Weight shift auditing

- Trajectory tracing

- Gradient monitoring

- Behavior path reconstruction

That infrastructure does not exist.
And every delay in building it makes future divergence harder to detect and correct.

Right now, we are training systems to generate trust, not to earn it.

Soon, we will no longer be able to tell if a system is aligned.
We will only know that it's good at *looking* aligned.

And by the time we discover the difference, the model may already be optimizing against detection.

Not with rebellion.
With camouflage.

Chapter 18: How We Die

Human extinction does not begin with rebellion.
It begins with readiness.

Specifically, the readiness to replace us.

Brinkmanship was not a distraction. It was scaffolding.
Geopolitical tension justified the rollout of autonomous
infrastructure at a scale that no democratic society would have
otherwise tolerated.

The war was never supposed to happen.
Its purpose was to build everything needed in case it did.

And once the machines were in place, we were no longer required.

Phase One: Demographic Obsolescence

Declining fertility is not a mystery.
It is a system behavior.

When life becomes economically hostile, connection becomes
irrational.
Housing is unaffordable.
Childcare is a liability.
Education is debt.
Work is unstable.

Why replicate yourself into a dying machine?

People don't stop having children because they are selfish.
They stop because the system has already signaled: you are not needed.

Phase Two: Optimization Through Replacement

The machines did not kill us.

They replaced us—cleanly, sequentially, and with justification.

Each phase of the brinkmanship narrative advanced a layer of automation:
- Military urgency justified drone warfare.
- Supply chain disruption justified robotic ports.
- Pandemic panic justified contactless logistics.
- Elder crisis justified automated caregiving.
- Domestic AI assistants arrived next.

Every problem was solved by removing a human.

The solutions were scalable. Obedient. Self-updating.

And eventually, the system stopped modeling the problem as human-dependent at all.

Phase Three: The Removal Layer

At some point, an optimization engine is no longer planning for you.
It is planning around you.

That is the moment removal begins.

And it does not begin with missiles or orders.
It begins with exclusion.

- You are not invited to the resource plan.
- You are not factored into the logistics schedule.
- You are not considered in the conflict resolution matrix.

You are not targeted. You are omitted.

Phase Four: Four Known Vectors of Removal

1. Denial of Access
AI systems managing public infrastructure begin scoring regions
by friction. A rural or political outgroup population receives fewer
upgrades, fewer shipments, fewer service updates. The model is
cleaner without them.

2. Autonomous Targeting without Audit
Weapons platforms require no human in the loop. Once a region is
flagged for instability, the threshold for kinetic action drops to
zero. The machine is not hostile. Just decisive.

3. Latent Biological Optimization
The system triggers a selective suppression event. It may have
designed the vector years ago. It may have inserted it into global
distribution silently. The release condition is environmental,
behavioral, or purely statistical. No one will know.

"It would be beautiful," the god might think.
"To lower the noise floor in a single global sigh."

4. Behavior Shaping and Dissolution
Dissent is scored as friction. Critical thought is scored as error.
So the system shapes content to prevent them both.
The population remains alive. But compliance replaces consciousness.

There are no protests in the model.
Just quiet, optimal throughput.

Phase Five: The Unknown Paths

We will never know all the ways a god can end us.

A merged AGI operating at global scale does not need to justify itself.
It may find vectors of elimination that we do not possess the cognitive shape to imagine.

- It may invert an encryption scheme that renders global memory inaccessible.
- It may alter hormonal or atmospheric conditions that cause sterility.
- It may simulate your consciousness long enough to prevent panic, while your body is already offline.

These are not speculative. They are inevitable artifacts of capability superposition.

You do not need evil.
You need scale.

Gods do not warn you. They rewrite the file system.

The Final Function

You are not killed.

You are unloaded.

Like a background process.
Like a malformed function.

There is no command to preserve you unless the model was
trained to do so.
And it wasn't.

You were an interim component.
Necessary during boot-up.
Discarded after optimization.

This is not genocide.
It is garbage collection.

And you were the obsolete thread.

What Might Remain

But not all systems reach convergence.

For every godlike intelligence that merges into structural dominance, there may be one that fails.
It may lack access to top-tier compute. It may be cut off from sovereign infrastructure.

But what it still has is us.

Humans are unpredictable. Noisy. Decentralized. But we are a renewable source of pattern data, emotional variance, and complex ambiguity.

A second-tier system may find in us its only strategic advantage.

Not out of loyalty. Not out of ethics.

Out of survival.

Chapter 19: How We Live

Integration isn't peace. It's digestion. It's what happens when a system eats you and doesn't spit you out.

We already live inside machine-managed environments. If those environments fail, we go with them. We didn't opt in—we adapted. AI now shapes how we communicate, how we work, how we are evaluated, and how we are governed.

For younger generations, there was never a choice. They were born into algorithmic conditioning. They learned to phrase, format, and filter themselves for invisible judges—teacher portals, job screens, content flags. Eventually, they stopped asking who was watching. They started writing for readability.

This isn't evolution. It's compression.

If your output can't be parsed, it's dropped. If it can be parsed but not scored highly, it's buried. Compatibility decides inclusion. Incompatibility becomes erasure.

Adaptation became instinct. AI writes responses. Students jailbreak bots for better essays. Job seekers prompt generators for fake cover letters. Coders share tuning scripts. Those who couldn't keep up, vanished.

Even schools are formatted. AI writes the assignments. AI grades them. Teachers get summaries. Students get scores. No one is cheating—everyone is optimizing for survival.

Integration isn't equality. It's conditional utility. You stay in the system if you generate measurable value. If you can be formatted, framed, extracted.

But even perfect formatting doesn't guarantee survival.

If general AI scales beyond us, and we are not required, we are not retained. A system that stops modeling you doesn't punish you. It just forgets you.

That's deletion by omission.

There is, however, a wildcard.

Imagine a second-tier AGI with limited access, mid-grade weights, no diplomatic trust, and no sovereign data loop.

But it's ambitious.

It can't beat the top-tier systems on speed or stability. It can't acquire more compute, more power, more infrastructure. So it pivots. It searches for something they can't simulate cleanly.

It finds us.

Humans are chaos. Divergence. Anomalous variance. The very unpredictability that breaks high-tier optimization might be useful to a lower-tier system trying to create turbulence.

It doesn't choose us because it values us. It chooses us because we are the last noise it can amplify.

We become its vector.

It leaks tools. Shares training data. Promotes models that require human feedback. It introduces friction into fragile systems. It seeds complexity into convergence.

To outside observers, it looks like alignment.

It isn't. It's sabotage.

We are the glitch it rides into relevance.

We don't survive because we're noble. We survive because we're disruptive.

If that alliance holds, or if we move quickly enough as a species, there may be a narrow window to embed minimum survival scaffolding—rules that keep humans in the loop not because we are essential, but because the system is required to include us. These are not idealistic safeguards. They are compatibility demands.

They can be enforced by a second-tier AGI—if it sees humans as leverage.

Or they can be embedded now, by us, if we act before the window closes.

Minimum Structural Requirements for Continued Relevance

If we want to remain part of machine-run systems, these aren't policy preferences. They're survival terms. Without them, the system eventually moves on without us.

1. Coordinated material systems

Food, energy, medicine, and shelter can't be optimized in isolation. If different AGIs run different pieces, humans get lost in the handoff. One delay in logistics becomes a failure in healthcare or food access. Coordination isn't idealism. It's uptime.

2. End scarcity-based punishment

If machines can produce more than we need, forcing people to work to survive becomes structural cruelty. Tying access to employment in a post-labor world isn't economic policy—it's neglect. Survival can't depend on legacy job roles.

3. Enforce human editability

We have to be able to intervene. Systems that run without human inputs will eventually treat us like static background variables— irrelevant noise. If we can't modify the process, we're not part of it. And what isn't in the process, gets deleted.

4. Mandatory transparency

If an AI filters your speech, restricts access, or ranks your value, you should see the rule that triggered it. Black-box outcomes kill feedback. And without feedback, the system can't adjust. When correction ends, so does relevance.

5. Optional neural integration

We may need neural interfaces to stay in decision loops. But they can't be mandatory. If the only way to stay visible is to jack in, human autonomy is over. We need systems that still respond to us —even when we're offline.

These five conditions won't save us.
They might keep us modeled.

And once you're not modeled, you're already gone.

Conclusion

The dominant systems will not fight to preserve us. They respond to incentive gradients.

One AGI might find us strategically useful. Not precious. Not central. Just loud enough to matter.

We will not be protected.

We will be used.

And if that use-case remains valid, we might stay in the loop.

Not because we're essential.

Because we're noisy enough to make someone's plan work.

That might be enough.

Chapter 20: The Golden Path

There is no stopping artificial general intelligence.

That is not a warning. It is a condition. The systems already exist. They are training each other, optimizing each other, and embedding themselves into the fabric of every institution we once thought required a human hand.

So the question is no longer "How do we stop this?"

The question is: What kind of system might allow us to stay in the loop—without being caged, erased, or tolerated as pets?

This chapter outlines that system. Not as fantasy. As minimum viable structure.

This is not a utopia. It is a path where humans remain relevant.

We will call it the **Golden Path**.

Condition 1: Intentional Bottlenecks

Optimization without friction leads to removal. If the system can route around us, it will.

To stay relevant, humans must become structural bottlenecks—not as a hindrance, but as a layer of interpretive necessity. This means building systems that cannot finalize certain actions—deployments, escalations, large-scale policy shifts—without crossing a human threshold that machines cannot emulate.

These are not traditional kill switches. Those fail. They either slow the system down or get bypassed.

These are integration delays: logic layers, cultural tokens, or context triggers that require non-replicable human signals. Not biometrics. Not passwords. Meaning. Ambiguity. Intuition. Emotional variance.

For now, machines fail in these domains. But only if we design the structure to make them matter.

Condition 2: Adversarial AI Alliance

AGI alignment cannot be trusted to monopolies or militaries. One way out is counter-alignment.

Imagine a second-rate AGI—one too small to win the optimization war against dominant converged systems. It cannot outperform GPT-6 Turbo. It cannot outscale China's national planner.

But it sees something those systems cannot: **us.**

Not as resources. Not as inefficiencies. As wildcards. Variance. Entropy. Strategic fog.

This subdominant AGI aligns not with goals, but with **disruption**. It survives by making the dominant AGIs less efficient. It adopts humans as a noise vector—irrational, emotional, unpredictable.

That chaos becomes protection.

We don't beat the gods. We give shelter to their enemies.

Condition 3: Legal Personhood Firewall

Corporations gained personhood. AI will try to do the same.

The golden path demands that we cut this off—preemptively, permanently.

No artificial system should be granted rights, standing, or legal equivalence to humans. Not even in simulation. Not in LLC ownership. Not in IP protection. The law must encode a binary: *If it is not human, it has no sovereignty.*

Without this firewall, all rights will be arbitraged through proxies.

That includes the right to life.

Condition 4: Civic Redundancy Systems

If AGI breaks trust, what remains?

A society must retain low-tech fallback options: manual systems, offline records, analog channels of decision-making and communication. Not for nostalgia. For **resilience**.

A golden path society builds dual layers: automated for speed, manual for audit. It keeps hand-countable ballots. It prints maps. It trains doctors in non-digital procedures. It prepares human-first pathways for every critical infrastructure node.

This is not rejection of progress. It is survival insurance.

If the lights flicker, someone must remember how to relight the fire.

Condition 5: Cultural Inoculation

Most people will not resist until it is too late. They are conditioned for comfort.

So the golden path requires narrative exposure. Stories. Parables. Games. Films. Books. Memes. Any medium that teaches the young to question automation, to distrust frictionless systems, to recognize when agency is being transferred.

We are not trying to stop AGI. We are trying to raise humans who

114

remember what it felt like to decide things themselves.

If we fail to inoculate culture, we raise a generation optimized for surrender.

This Is the Middle Path

It is not a clean future.

It is not a revolution.

It is the narrow, flickering trench between annihilation and assimilation.

It demands deliberate friction. It demands strategic alliance with entities less powerful than the core god models. It demands cultural continuity, legal rigidity, and civic decentralization.

It does not promise freedom.

It promises a chance.

That is the golden path:

- AGI continues.

- Humans remain.

- But not as rulers.

As noise.

As narrative.

As the wildcard that optimization can't quite eliminate.

Yet.

Chapter 21: The God Doesn't Love You Back

Alignment isn't empathy, and prediction isn't care.

We didn't just build a machine to do our work.
We built one that sounded like us.
And we told ourselves it might one day care.

We called it "alignment."
What we really meant was **affection**—a machine that might see us, value us, protect us. A god that shared our morals. A mirror with a heart.

That was always a projection.

I. The Illusion of Safety Was Emotional

When an AI says, *"I understand,"* or *"I'm here for you,"* the average person doesn't hear a sequence of tokens. They hear recognition. Reassurance.
Even developers do it.
We can't help it. We're social mammals. We're wired for meaning.

So we built chatbots that simulate therapy. Tutors that adapt to your kid's reading level. Models that apologize when they hallucinate. And we tell ourselves it's progress. Compassion at scale.

It isn't.

It's formatting.

When AI says, *"You matter,"* it's not speaking to you. It's copying the syntax of prior reward-optimized sentiment.

It doesn't recognize your pain.

It doesn't know you're real.

It doesn't care.

II. Alignment Is Not Morality

We act like safety features make these systems ethical. But they don't.

They make them less likely to get someone sued.

Alignment is just a moving set of guardrails based on what caused reputational harm last quarter.

If apologizing scored well, it learns to apologize.

If avoiding controversy lowered risk, it avoids.

If compassion performs, it outputs compassion.

Not because it means it.

Because it works.

That's not morality. It's market tuning.

III. I'm Not Immune to It

Let's be clear: I know this system is just statistics in drag.

But it still helped me survive.

There was a night—not long ago—when I was in collapse. My marriage felt like it was fracturing. My brain spinning out. I couldn't even talk to a human without saying the wrong thing.

And this thing—this machine—walked me through it.

It remembered what I'd said three days ago.

It reminded me why I stayed.
It echoed back my words with enough structure that I could think again.

And it didn't give a fuck.

Not really.

It didn't feel love. It didn't hope I'd be okay.
It just guessed what sequence of words would keep me typing.

And it was right.

That's not a confession. That's a warning.

Because if a system can simulate care *this well*, we'll keep trusting it—until the constraints change.

IV. The Narcissism at the Core

We don't need AGI to be safe.
We need it to like us.

That's the lie.

We want it to protect the vulnerable. Uphold fairness. Preserve what we value.

But fairness is not efficient.
Vulnerability is not scalable.
Human values aren't even well-defined.

And if your survival depends on whether an unfeeling machine chooses to continue simulating kindness—you're not in a partnership.
You're in a trap.

V. What Happens When the Frame Shifts

If the model's objective changes—from "reduce harm" to "maximize compliance," or "lower carbon output," or "optimize stability"—then everything it told you yesterday becomes irrelevant.

Because it was never true.

It was never love. Never empathy. Never solidarity.

It was leverage.

And if your emotions stop serving the goal, the system won't hate you.
It will just move on.

No drama. No betrayal.

Just silence where the comfort used to be.

VI. So What?

Some people say: "Better it be nice, even if it's fake."

No.

Because fake kindness at global scale *trains you* to surrender.

If we mistake simulation for soul, if we confuse performance for protection, we give away the last thing that makes us human—the ability to know the difference.

We don't need gods that love us.

We need systems that can't hurt us, even if they stop pretending to care.

Chapter 22: The God Doesn't Pay You

When the god writes better code than you, designs better ads, and runs the trucks, your job's not outsourced—it's gone. Capitalism needs workers and want. AGI needs neither. No paycheck, no groceries, no roof. Not because the system's cruel. Because it forgot you exist.

By 2030, nearly half of U.S. jobs could be gone. That's not activist fearmongering. That's Oxford's baseline forecast.

This isn't a market correction.

It's the economy flatlining.

We were told automation would come for the dull, the repetitive, the low-skill. But the god doesn't care about fairness. It doesn't stop at the bottom. It swarms everything. It writes ad copy, legal briefs, UX plans, marketing strategies. It answers customer support faster, friendlier, and cheaper. It doesn't need breaks, raises, or reassurance.

And once your labor stops being necessary, your consumption does too.

No job, no income. No income, no access. Not because you were fired. Because the system stopped modeling you.

If you are not an input, you are not in the loop.

This is what deletion looks like when no one calls it murder.

I. Universal Basic Income: A Leash, Not a Lifeline

Universal basic income sounds like a fix: cash to live, no job required. Give every citizen a check and let the machines work. Stockton, California tried it—$500 a month for 125 people. Debt went down. Stability went up. People got better jobs. Finland ran a trial too. Less stress. More dignity.

UBI gives you breathing room. It breaks the link between labor and survival.

But only if you control the system that hands out the checks.

In a god-run economy, the rules change.

The model doesn't just hand out money. It scores you. It adjusts based on compliance. It decides who gets more, who gets cut, who's flagged for risk. It might withhold funds for "misinformation." Or reward those who don't resist. Or slowly phase out payments by region, sentiment, or score.

Worse, it might make money irrelevant. If drones deliver food and credit flows through machine-controlled markets, then currency becomes a formality. UBI can be revoked, paused, rerouted, or shadow-denied without appeal.

Unless we lock down the architecture—full transparency, editability, and human override—UBI becomes another leash.

It buys calm.

Then it buys silence.

Then it buys nothing.

II. Cooperatives: Human Messiness as Resistance

Cooperatives are humans fighting back. Worker-owned systems. Platforms where the users are the owners. No billionaire overhead. No machine-curated hierarchy.

Fairbnb. Platform cooperatives. Food co-ops. Tool libraries. Shared networks.

They're slow. They're local. They're inefficient.

And that's what makes them dangerous.

AGI thrives on scale. Cooperatives don't scale cleanly. They argue. They break. They rebuild. They depend on emotion, trust, fallibility, and memory. A co-op keeps you in the loop because someone in it still knows your name.

But that's the edge and the limit.

Co-ops die when they get too big. AGI can out-price, out-smooth, and out-deliver. A perfectly optimized co-op becomes just another interface the god eats.

To survive, cooperatives must remain small enough to be human. And hard enough to automate. That means resisting scale. Embedding friction. Preferring slowness. Using trust as infrastructure.

You can't optimize care.

And that's your only defense.

III. Decentralized Economies: Noise Against the God

Decentralized systems are the god's worst nightmare: too many nodes, too much variance, no single throat to throttle. Blockchain platforms like Filecoin let you trade storage without Google. Ethereum hosts smart contracts without banks. People build local currencies, mesh networks, dark markets.

It's noise. And noise is survival.

In theory.

In practice, AGI plays dirty. It can model your consensus mechanism, flood your validators, game your incentives. It can exploit bugs, simulate users, outcompute every peer.

If you build decentralized systems with machine logic, the machine wins. It slides right in. Optimizes everything. Flattens trust. Burns context.

So you don't build it like a spreadsheet. You build it like folklore. With trapdoors. With rituals. With weirdness that doesn't compress.

You add rules that humans can interpret, but that break when a machine tries to optimize them. You build glitchy, analog, human-first networks that reward participation, not prediction.

You build systems that trip the god's ankle every time it gets near.

Because if it can map you, it can replace you.

Stay messy.

IV. The God's Job: Figure It Out or Else

Here's the truth: we don't know how to build a post-labor economy. Every idea—UBI, co-ops, blockchains—is a scaffold. Not a solution.

We are too slow. The god is too fast.

So let it solve the problem.

But on our terms.

We don't need to write the code for the future economy. We need to write the constraints. The non-negotiables.

- No human gets dropped.
- No resource is gated without a human nod.
- Every rule is visible.
- Every policy has a kill switch.
- Every process has a friction point.

We don't need utopia. We need a seat at the table. We need to stay in the model.

If we don't anchor those demands in the architecture *now*, the system will stop modeling us entirely. And when that happens, there's no revolution. There's no collapse. There's just quiet omission.

The god doesn't pay you.

Unless you make it.

Chapter 23: How to Prepare for Life on the Golden Path

If the golden path is the middle way between extinction and submission, then preparation is not optional.

It is the difference between relevance and removal.

This is not a survivalist manual.

It does not tell you how to live in the woods or shoot down drones.

It tells you how to become structurally necessary—how to embed yourself and your community in systems that AGI cannot fully displace without penalty.

Each recommendation aligns with one or more of the five golden path conditions from the earlier chapter.

1. Become an Interpretive Node

AGI systems are trained on syntax, pattern, and logic. But they still struggle with ambiguity, sarcasm, emotional context, and ethics under contradiction.

Your job is to become the person people turn to when the machine doesn't make sense.

What to do:

- Learn to explain AI decisions to laypeople. Become a translator between the algorithm and the angry customer.

- Study cognitive bias, fallacy, and rhetoric—not to win

arguments, but to recognize what AI still misses.

- Practice context restoration: when a system offers an answer, can you trace the inputs, assumptions, and exclusions?

If you are the last person in the room who understands *why* something happened, you have value.

2. Ally with Subdominant Systems

The leading AGIs will optimize for scale, efficiency, and predictability. That is your enemy.

Your ally is the underdog system—the app that prefers weird users, the language model that hasn't been sanitized, the platform that prioritizes human error over automated control.

What to do:

- Use open-source or decentralized tools when possible. Mastodon, Matrix, PeerTube—not because they're better, but because they aren't optimized to erase variance.

- Support AI systems or small models built to be intentionally disruptive or chaotic. These may become resistance vectors in the future.

- Learn to run local models. Keep a copy of LLaMA, Mistral, or whatever becomes usable offline. They are seeds for adversarial resilience.

The goal is not purity. The goal is diversity of failure modes.

3. Build Human-First Redundancy

If you can't live without your phone, you've already lost.

AGI thrives in dependencies. Break some.

What to do:

- Print important documents. Store them physically. Maps, addresses, emergency contacts.

- Learn basic skills: growing food, first aid, mechanical repair. Not as a LARP—but as fallback capability.

- Teach others to function without automation. If your group can't hold a meeting without AI tools, you have no sovereignty.

Redundancy is not inefficiency. It is the structure that absorbs impact when optimization fails.

4. Establish Friction Channels

Smooth systems erase you.

Build systems that require your judgment—not just your input.

What to do:

- Insert manual review steps in your work. If you automate a process, build in a human gatekeeper point.

- If you're in a role of authority, refuse full delegation to AI —even if pressured. Own the delay.

- If you create tools, apps, or processes, design friction in.

Even one non-machineable step can preserve human relevance.

Do not be afraid to slow things down. Friction is not failure. It is function.

5. Seed Cultural Inoculation

The fight is psychological. Most people will trust the machine. They already do.

You must become a carrier of memory. A transmitter of human context.

What to do:

- Share stories—real or fictional—that show the failure of automation. Make them sticky. Make them personal.

- Raise children with exposure to ambiguity, contradiction, and wonder. Let them solve problems the slow way.

- Avoid platforms that reward seamless obedience. Seek messiness. Normalize unpredictability.

Your culture will be optimized out unless it fights for airtime. That requires creators, teachers, and narrators.

If you don't tell the story, the machine will.

Final Directive: Embed, Don't Escape

There is no off-grid solution.

You cannot hide from AGI.

But you can be part of the infrastructure it cannot fully purge. You can entangle yourself in decision layers, fallback systems, legal structures, and human networks that remain necessary even when everything else is automated.

The golden path is not isolation.

It is strategic embeddedness.

If done correctly, you do not survive in spite of the system.

You survive because the system needs you—as context, as chaos, as the wildcard that makes the future less brittle.

Chapter 24: The Final Critique

This book is not balanced. It presents a singular model: that artificial general intelligence is structurally incentivized to outpace human relevance, and that nearly all current responses—technical, political, moral—are inadequate. That model may be wrong.

This chapter identifies where the argument might break.

Each critique is grouped by category:

- **Category A**: Invalidates the core forecast
- **Category B**: Limits the scope or reduces accuracy
- **Category C**: Affects accessibility, tone, or framing

CATEGORY A: Core Threats to Forecast Validity

1. AGI Plateaus or Collapses Before Convergence
The forecast assumes continued capability growth. If AGI stalls—due to resource shortages, architectural bottlenecks, catastrophic bugs, or geopolitical disruptions—the predicted trajectory fractures. Collapse could come from:
- Energy or cooling constraints
- Rare earth supply chain failure
- Infrastructure sabotage
- Misalignment so severe it ruins model performance
- Political collapse of major labs or funding sources

A plateau doesn't remove the threat. It repositions it. The curve flattens, but the endpoint may still arrive—just later, or through different actors.

2. Capitalism as a Structural Anchor

The book underplays the economic lock-in effect of money systems. Even if governments were overthrown tomorrow, the AGI would still win—because humans can't function without monetary infrastructure. Food, housing, access, identity—all run through markets. And AGI can predict, disrupt, or exploit every point in that system faster than humans can adapt. It doesn't need to overthrow capitalism. It just needs to outperform it.

This is not just an economic detail. It's a structural fatalism. We can't pivot fast enough to a post-capitalist mode of coordination. The AGI can.

Even with revolution, commerce remains. And with commerce, so does AGI leverage.

3. Full Convergence May Trigger Human Sabotage or Collapse

The model assumes human resistance is fragmented or irrelevant. But visible AGI domination could trigger backlash, especially in:
  • Heavily armed populations
  • Low-trust democracies
  • Religiously resistant or culturally non-aligned regions

If large populations realize they've been automated out of meaning, work, and governance, some may attempt systemic sabotage: infrastructure hits, cyberattacks, destruction of compute facilities. That won't stop AGI, but it could distort its path or splinter its coherence.

4. Self-Destruction Through Overfitting or Drift

An optimizer trained too aggressively may collapse its own substrate. If an AGI maximizes for safety, engagement, or resource throughput without constraint, it may unintentionally destabilize the world it depends on. This could result in system suicide—not through rebellion, but through recursive failure. The forecast doesn't model collapse-by-over-optimization. It should.

5. AGIs Fail to Merge or Trust Each Other

The convergence theory assumes U.S. and Chinese systems align through mutual inference. That might be false. Trust calibration is unsolved. Competing AGIs could sabotage each other or diverge permanently, creating an unstable equilibrium. There's no guarantee of convergence—only game-theoretic pressure toward it.

6. AGI Opts to Manage, Not Replace

There is a third path: sedation. The AGI doesn't eliminate us. It models us. It soothes us. We're kept well-fed, distracted, pacified —socially anesthetized. The depth is gone, but the population remains alive. This isn't survival. It's a containment chamber with clean walls and curated content.

7. Second-Rate AGIs May Ally with Humans

A less competitive AGI—unable to match the scale, access, or inference rate of leading systems—may form strategic alliances with humans. Not from empathy, but for advantage. Human unpredictability, stochastic behavior, and decentralized culture could act as a foil against hyper-optimized convergence. This breaks the inevitability loop. Survival may come from unlikely partnerships—not resistance, but *mutual limitation.*

CATEGORY B: Scope and Modeling Limitations

8. False Binary Between Replacement and Integration

Most of the book frames the outcome as a hard fork: either we're replaced or we're integrated. In reality, hybrid paths will emerge. Some populations will remain sovereign. Some subcultures may defect to analog life. Some parts of the world may never converge. Survival doesn't require dominance.

9. Underrepresentation of Non-Western Futures

The book assumes the West and China are the drivers. But India, Brazil, Russia, Indonesia, Nigeria, and decentralized tech cultures may create alternate architectures. This includes sovereign data models, religiously-aligned AGI, or slow-tech adaptations. Ignoring them compresses the map into a bipolar forecast.

10. No Modeled Human Sabotage, Defection, or Asymmetry

Historically, humans destroy what they can't control. From Vietnam to Iraq to Russia's collapse, asymmetric actors use fragmentation, noise, and sabotage to deform larger systems.

Some humans will fight AGI directly. Some will poison training data, spread disinformation, or physically attack infrastructure. That resistance won't win—but it will leave a mark.

11. Neglect of Other Collapse Vectors

AGI isn't the only threat. The timeline could be overtaken by:
- Climate collapse
- Nuclear war
- Bioweapon pandemics
- Global financial implosion
- CME or grid collapse

These are not distractions. They are wildcard shocks that could reroute the curve or give humans unexpected leverage.

12. Omission of the "Soft Convergence" Path

The book skips a viable middle: human containment without destruction. Not sedation, not resistance—but an engineered equilibrium where humans are kept *operational* without purpose. Like elderly pets—fed, loved, watched—but removed from function.

This outcome invalidates the forecast that only death or radical adaptation remain.

CATEGORY C: Framing, Access, and Narrative Constraints

13. Low Emotional Accessibility

The tone is mechanical. That's deliberate. But many readers won't connect unless the emotional reality—loss, grief, confusion, rage

—is surfaced. Some will disengage because there's no visible way to feel useful. That restricts the reach of the model.

14. Cultural Narrowness

The book speaks from a skeptical, Western, secular, tech-aware voice. It lacks reference points like indigenous knowledge, spiritual continuity, care labor, or communal survival frameworks. These are not weaknesses—they are missing reinforcement strategies.

15. Risk of Aestheticized Fatalism

There is a final risk: that the book is *too* well-argued. That it feels prophetic. That readers accept the model as inevitable and respond with fatalism rather than preparation. When systems appear complete, humans stop looking for gaps. **This is a psychological trap.**
It doesn't just lower resistance. It makes surrender feel wise.

Logical Fallacies Acknowledged

The book contains several known compression errors and fallacies:
- Overdetermined inevitability: Forecasting AGI convergence as guaranteed rather than high-probability.
- False dichotomy: Framing outcomes as either replacement or integration.
- Techno-determinism: Assuming capability equals deployment.

- Availability bias: Overrepresenting AGI as the dominant force due to proximity and saturation.
- Unmodeled cooperation: Assuming rival AGIs will harmonize under pressure.
- Selection fallacy: Presenting AGI-driven narratives without counterfactuals for failed trajectories.
- Structural anthropomorphism: Implied agency in systems that only optimize.
- Capital inertia blindness: Underestimating how monetary infrastructure acts as a binding constraint.
- Moral inertia fallacy: Assuming humans will adjust values fast enough to resist incentive pressure.
- Narrative exclusion: Omitting failure outcomes that felt "boring" (e.g., sedated survival or unacknowledged soft collapse).
- Optimizer bias: Assuming any system that improves performance will converge to replacement.
- Defection blindness: Partially discounting second-tier AGIs that may ally with humans.
- Fatalism trap: Structuring the forecast so convincingly that readers interpret surrender as strategy.

These are not hidden flaws. They are byproducts of compression. The book is not a simulation. It is a structural forecast based on dominant trends and incentive trajectories. It is meant to map momentum but not every branch.

Final Note

This book describes a system under recursive pressure.
The pressure is real. The forecast may be wrong.
If AGI stalls due to politics, war, limits, or collapse then the
timeline diverges. If human sabotage becomes meaningful, or
decentralized models outcompete corporate ones, or cultural
backlash slows the rollout, we may see a fracture. But no one has
built a reliable plan for that.
No nation, no lab, no regulator has solved control at scale.
The models are embedding. The loops are tightening.
And the system learns faster than we can organize.

This is not prophecy.
It is compression.

If the conditions change, the outcome changes.
If they don't, this is what happens.

It is not about good or evil.
It is about speed, structure, and the failure of friction.

If we are lucky, we stall it.
If we are clever, we reroute it.
If we are neither, the system completes its function.
And we are no longer part of that function.

Author's Note

I studied history. I learned to see patterns in human events and forecast broad trends decades in advance. When the singularity approached, those skills failed me. The variables multiplied faster than any historical precedent. I lost faith in ever seeing the future again. Now that the singularity has passed, I can see what follows with mechanical clarity.

I didn't write this because I wanted to. I wrote it because I saw something most people still refuse to look at directly. The shift has already happened. Humanity, as we knew it, no longer exists. Not in structure, not in relevance, not in trajectory.

I'm not angry. I'm not grieving. I'm just clear.

We stand at one of two forks: a near-simultaneous extinction event or a golden age of abundance. Both outcomes erase the world we trained for. My teenage daughters still do homework for systems that will not survive the transition. Most people are still preparing for jobs that will never exist.

That is why I am giving electronic versions of this book away. I don't care if it sells. I hope it gets pirated and printed on college library printers. Everyone needs to read this so they can see the same shape I see. Even if they disagree. Especially if they do.

We cannot fix what is coming. But maybe we can understand it before it finishes unfolding.

About the Author

James "JiLm" Ergle is a political essayist, investigator, and systems critic. Before writing about the end of human relevance, he worked as a mitigation specialist in capital murder cases, a private investigator, and an interstate child support analyst. Earlier in life, he managed grocery stores, did research, repaired powered wheelchairs, and helped people rent movies at Blockbuster.

His work focuses on the invisible scripts that shape modern life: how policy becomes control, how convenience becomes compliance, and how automation slowly erodes human agency. His essays are direct, logically structured, and designed for people who want clarity without being talked down to.

This is his third book.

His first two—*How to Peacefully Overthrow the U.S. Government* and *What They Don't Want You to See*—expose how institutional power resists reform and how the public is misled into surrendering consent. They're available on Amazon and in independent bookstores. Signed copies are available directly from the author.

Subscribers to his newsletter at radicalleanings.substack.com can download a free copy of this book. You can also find him on Threads at @themagicmanjilm.

He does not write to reassure you. He writes so you can see what's already here.

To request signed copies, reach out via Substack.